Internet Guide For Maintenance Management

PROBLEMS   SOLUTIONS

Joel Levitt

Library of Congress Cataloging-in-Publication Data

Levitt, Joel, 1952-
    Internet guide for maintenance managers / by Joel Levitt
        p.192 6x9
    Includes illustrations and index.
    ISBN 0-8311-3081-4
    1. Plant maintenance--Computer network resources. 2. Internet
    (Computer network) I. Title.
    TS192.L468   1998
    025.06 6582 02--dc21   98-42765
                                                    CIP

INDUSTRIAL PRESS, INC.
200 Madison Avenue
New York, NY 10016-4078

FIRST EDITION

10 9 8 7 6 5 4 3 2 1

Dear Reader,

Thank you for purchasing this book.

As you know, the Internet changes and grows on a daily basis. Some of the sites listed in this book, despite our best efforts, may be given new addresses or even terminated by the time you try them. Besides, some of the examples in this book may change by the time you view them "live" on the Internet.

For your reference, we will maintain an up-to-date summary of links at two locations: http://www.maintrainer.com (under the maintenance resources button) and http://www.industrialpress.com (under the related maintenance sites button).

I sincerely hope that you will give each site a try, including those we thought somewhat skimpy on content when this book went to press. Many will have been upgraded and improved by the time you take a look.

Finally, a note to the thousands of surfers, vendors and friends of the maintenance community: Please send the URL of any site you think should be considered for the next edition to: jdl@maintrainer.com. Don't be shy.

Thank you and I hope you enjoy the trip.

Joel Levitt

DEDICATION

This book is dedicated to the memory of two friends.

Jay Butler was a leading thinker, designer, teacher and mentor in the field of maintenance management. He introduced me to the field, paved the way for my involvement and education. His designs, constructs, and priorities for maintenance systems still dominate the field. He passed away just as the Internet started to have an impact on how business was conducted. He would have loved the rough and tumble world of the Internet. I miss his wry wit and totally unique way of looking at maintenance management problems. I also miss his friendship.

Maryanne Cavanaugh was my wife's younger sister. She passed away tragically after a long illness. She taught us all many lessons, two of which seem particularly important to me. If life is a game, Maryanne played the game full out, to the end. Well after the 'professionals' said her time had come she continued to live fully. She visited three coasts, played with her nieces and nephews, swam with dolphins, and visited shrines as an expression of life. The other lesson was her faith. With her faith she faced her illness with an inexhaustible positive attitude. Its hard to imagine her gone, I miss her.

Table of Contents

INTRODUCTION TO THE INTERNET FOR MAINTENANCE PROFESSIONALS

The Internet is everywhere. You can't read a newspaper or see a TV broadcast today without hearing about the Internet. Yet, how useful is it today for busy maintenance professionals? Robert Baldwin, the editor of the magazine Maintenance Technology, has written regularly about the Internet and maintenance management. In his February 1998 "Uptime", column he comments:

> Where is the good stuff the information that can help you understand technology or business issues related to a company's offerings? A few sites provide some crumbs by posting their company's newsletter and application stories from customers, but that's about all."

> Where are the white papers, maintenance tips, rules of thumb, glossaries, do's and don'ts, frequently asked questions, and installation checklists?

> The lack of good maintenance stuff on supplier Web sites reminds me of that delightful 1984 television commercial where Clara Peller looked up at the server behind the fast food counter and asked: "Where's the beef?"

The Internet is getting better for maintenance users on a daily basis. It is changing rapidly because the customers are becoming more demanding of the goodies that Bob Baldwin asked for in early 1998. You can contact him and give him your opinion by E-mail at editors@mt-online.com.

Internet pioneers were wowed by the technology, brochureware (catalogs on the web) and the ability to send a message and get a response.

Today, maintenance professionals active on the Internet are more sophisticated and need more return on their time invested than even just a year ago. We have good reason to think that even greater change will come in the next years.

Maintenance professionals who are not using the Internet have another problem. They fear the world will pass them by if they don't start using the Internet. The funny thing about the Internet is its multiple personalities. Is easy to use (the learning curve is about an hour after you are hooked up, assuming you can already use a computer), is mostly filled with trivia; yet at the same time, it is the answer to the prayer of maintenance professionals for around-the-clock access to information. Even as teenagers are chatting about the newest music, a researcher is accessing breakthrough up-to-the-minute research on the human gnome project.

The Internet is changing the way we communicate. E-mail alone has brought together families across the globe, authors writing books on three continents (at the same time), and presidents of companies with their employees. The change is radical, like the changes brought to maintenance by the fax machine or the computer itself. The maintenance field is in flux because of the new capabilities available on the Internet. Yet the Internet is being used by only 5 to 10% of maintenance professionals.

WHAT'S IN IT FOR MAINTENANCE PROFESSIONALS? FIFTEEN USES OF THE INTERNET:

The Internet is being used by the maintenance profession in many ways. Since the capabilities are on-line, they are available around the clock, 365 days a year. Usually the servers (computers where the information is located that are connected directly to the Internet) are available except when they are being backed-up or serviced.

1. The biggest use of the Internet is carrying messages or E-mail. According to one study, 80% of the business uses of the Internet were for E-mail.

2. Finding vendors of everything from valves to engineering services. Companies can make their latest catalogs available as soon as they are complete. It is much less expensive to provide catalogs on-line than to print them. On-line catalogs save your shelf space and trees too! Because of the increasing cost of paper, expect to see a push for on-line catalogs. Since storage on the computers is inexpensive, a huge volume of information can be made available such as complete technical specifications, photographs, video clips, audio descriptions, or drawings. All of the information is just a click away.

Locating vendors is the second most popular use for the Internet after E-mail. The driving force is the advertising budget. The fee for an entire Web site for a year is comparable in cost to a single full-page ad in a leading maintenance magazine.

3. Technical bulletins: Information about the latest technical problem and fixes can be available minutes after the vendor's engineers decide to put it on-line. No longer is there a weeks-to-months lead-time to publish and mail the bulletins. The software vendors are light-years ahead of everyone else in this area and give a higher level of support at a lower cost through this method (see item ten below on software bug fix, software distribution).

4. Drawings, field modifications, and manuals: The same way that you can be updated by technical bulletins, you can view manuals and download drawings, (Download means to copy a file from the server computer, to your computer). The file can be a manual, a drawing, just about anything. Wouldn't that be great at 3 AM when you can't find the wiring diagram? Also, field modifications can be fed back to the Original Equipment Manufacturer (OEM) engineering departments if that is appropriate.

5. Parts information, parts purchasing, reducing the cost of acquisition: Some sites allow you to look up part numbers from exploded drawings. You can move your mouse cursor to the part and then drag it's number to an order form. Once you add your Purchase Order number and ship-to address, you have placed an order.

6. Commerce: This use is an expansion of the idea above. You can currently shop for many MRO items from storefronts on the Internet. Major industrial distributors such as Grainger and McMaster-Carr have a large presence on the World Wide Web. These storefronts currently cover all types of consumer goods and a few offer tools, maintenance supplies, uniforms and other items. Encryption (a fancy way to scramble transmissions to make them hard to intercept and make sense of) is becoming widespread to allow high security for credit card numbers and bank information.

7. FAQ (frequently asked questions): Every field and every piece of equipment has FAQ's. These types of basic questions take up most of the time of the telephone support department. Novices, new customers or customers new to a specific product can read the FAQ file. Many of the larger FAQ files have search engines that allow the user to make specific inquiries. FAQ's are on-line and available 24 hours a day, when you, the new user, have a question.

8. Technical help: Technical help is one of the greatest uses of the Internet. You can ask questions of the vendor's technical departments and get answers back to solve your problems. Technical departments develop a menu of canned E-mails that provide solutions to common problems and can be sent immediately. The technician can then spend time on the more uncommon or complicated problems.

9. Locating used equipment and parts: There are many classified ad sites where companies and individuals can buy, sell and trade equipment. For example, a local manufacturer buys and sells punch presses completely on the net.

10. Software changes: Almost all-major vendors of software allow you access to the latest versions of their software. You can visit their site and can initiate a download of the latest version. Also, software that you want to sample is available to download..

11. Directories of installers and vendors: When you are looking for vendors or installers, you can ask members of a newsgroup related to the topic, make an electronic query from a home page or send an E-mail to the companies' postmaster or webmaster.

12. Access to libraries: Many university libraries and information databases are available on-line. The Library of Congress is putting its enormous library on-line. Another group is making the complete texts of great books available via downloading.

13. User groups: Do you own a CMMS (computerized maintenance management system) and want to talk to others using the same system? Many user groups are going on-line as newsgroups (see below). Here you can read others' comments about the software, ask questions of the whole group, get help, and gripe to your heart's content.

14. Newsgroups: These are groups that are bound by a common love, hate, interest, or membership. By mid-1998 there were almost 35,000 newsgroups on the Internet with new ones starting everyday. Groups range from people who collect stamps or love anagrams, to people who hate politicians, or fast food.

15. Killing time: If you have an hour or more to spare, the Internet can be more fun than TV and a lot less predictable. In some homes, web surfing has almost replaced channel surfing.

MAJOR CAPABILITIES OF THE INTERNET

E-mail is one of the most used and most powerful parts of the Internet. It links the entire world together and enables researchers, business people, and even elementary school kids to send messages worldwide. There are no extra charges beyond the local phone call to your ISP (Internet Service Provider). A recent survey showed that over half of the users of the Internet just used E-mail. Related to newsgroups (described below) are E-mail mailing lists. Mailing lists are lists of people's E-mail addresses. The people on the list share some common interest. The postings (comments from other subscribers) get sent to their E-mail box. You send a message to the whole group by sending it to a special E-mail address, which resends the message to the whole list. Over 71,000 lists were identified by mid-1998.

The World Wide Web (WWW, or just the Web) is where the explosive growth is taking place on the Internet. Any organization that can afford $100 per month can have a home page on the WWW. All of the addresses that start with http://www. are World Wide Web sites. The World Wide Web was designed to allow graphic transfers of information. Among of the most powerful aspects of the WWW is the ability to hot link (hyperlink) to related sites on the WWW. The hyperlink capability allows you to 'surf the net'. Hyperlinking is also the area of great interest to the maintenance community.

FTP (file transfer protocol) was one of the early ARPANet (predecessor to the Internet) capabilities. Using FTP you can visit thousands of computers and copy files to your own computer. These files could be weather maps, programs to solve engineering problems, games, electronic books, bibliographies, or just about anything else. FTP sites allow access to public directories that you can browse (although you have to use the next capability, Telnet, to browse). In the newest browsers, FTP sites are indistinguishable from Web sites. The interfaces look the same. The mechanics of the FTP are handled entirely by the browser.

Telnet (the other original capability) allows you to go to a remote computer and act like you are directly connected. You can browse the directory, run programs, or can do almost anything a local person can do. Telnet and FTP were early great applications that made the whole idea of Internet computing powerful and useful. Telnet is common in scientific sites such as supercomputer centers and less common in business. Dr. Mark Goldstein predicts that in the near future some version of Telnet will become very popular when you will use remote computers across the Internet to process and store your maintenance data.

Newsgroups are people interested in particular topics like real estate investing, wine tasting, presidential politics.

Search sites are essential in an entity growing as fast as the Internet. These sites are the card catalogs of the Internet. Most of them include robot programs called spiders that periodically search all of the Internet sites for key words and ideas. The search engine's server creates an index file from the spider's walk through the Web.

For example if you are interested in albino reptiles, you enter the words "albino reptile" into the engine's search screen. After pushing the search button the program searches its indexes for any sites that mention both the two words. All sites that have a match for both words are listed in order of relevance (the number of entrees that match or are close to your request).

WHERE IT ALL CAME FROM

The Internet runs on a technique called packet switching. This communication technique allows computers to have several conversations going on each communications channel at the same time. Packet switching made electronic mail or E-mail, telnet and FTP possible. By itself, Email is something of a revolution, offering users the ability to send detailed letters at the speed of a phone call.

As this grew, some high school and college students developed a way to use it to conduct online conferences. These started as science-oriented discussions. But they soon branched out into virtually every other field, as people realized the power of being able to "talk" to hundreds, or even thousands, of people around the country.

The Internet is a completely decentralized network; its communications backbones (high-speed communications lines) are spread all over the world. It is designed to survive even if major sections are removed or damaged. Originally the Internet and its predecessors ran exclusively on text with no pictures, color or sound.

Please turn to the Appendix section titled, History of the Internet for a blow by blow description of the first 30 years of the Internet.

MECHANICS OF THE INTERNET

The mechanics of getting on the Internet have changed dramatically over the last few years. Until 1996 the Internet was a do-it-yourself proposition. Each upgrade of software makes the Internet simpler and more transparent. The latest Windows product makes the Internet an icon on your screen. One click, and you're surfin'.

HOW TO GET ON THE INTERNET FROM HOME OR FROM A SMALL COMPANY

If you are trying to connect to the Internet at home or if your office has no gateway, then you will have to take three simple steps.

1. You will need some hardware. The hardware will include a Windows/Intel computer (sometimes called a Wintel computer), Apple computer or other advanced system with a modem. The modem needs to be hooked-up to a working phone line. Modems can share lines with phones or faxes, but if you have the call-waiting feature it should be turned off whenever you use the modem. Otherwise, the incoming call tone will kick you off the Internet.

2. You will need an ISP (Internet Service Provider). There are thousands of ISPs throughout the country. They include the long distance companies, regional phone companies and many other local and national providers. Specialized service providers such as America Online are good choices for many users. Consult the business section of your local newspaper or your local yellow pages for a wide variety of ISPs.

3. You will also need software to dial-up and connect you to the ISP. You need a second program to display the Internet and allow you to surf (this second program is called a browser) The software is free with your operating system or from your ISP

The ISP allows you to dial into its computer (called a server). The server is connected to the Internet, 24 hours a day with high-speed communication lines. Many people such as you share the ISP's high-speed lines to the Internet. In the major markets the flat fee is less than $25 per month.

While there are thousands of service providers, I would start with the larger ones such as AT&T, Sprint, MCI, most of the Baby Bells (regional phone companies), or AOL. AOL is unique in the group because it offers additional services over and above the Internet. (AOL is the provider that constantly sends out free "try me for 50 hours free" CDs.

HOW TO GET ON THE INTERNET IN A LARGE COMPANY

Your organization probably already has an Internet account (called either a PPP or SLIP account) or its own local server. The first people to ask are your friendly IT (Information Technology) department. The Internet might be available through a gateway available on your office PC LAN (Local Area Network). In a few cases, the Internet is available simply by turning on the Internet option.

NOW THAT YOU'RE CONNECTED WHAT NEXT?
WHERE IS ALL THIS SURFING I HEAR ABOUT?

You sign up with a service provider and it sends you a disk with all of the software. Follow the directions and the set up is completed in about twenty minutes. The process to set up an Internet account is pretty simple and highly automated.

With AT&T WorldNet, my off-brand modem didn't work with the software. A WorldNet technician on the 800 service line suggested an alternate modem. His suggestion worked on the first try. The entire process, with waiting on hold, only took forty minutes.

To start an Internet session, just click (or double click, depending on the computer) the Internet icon on you're screen. In computer books the screen is called the desktop (sort of a confusing idea). So you might see, double click the Internet icon on your desktop. If you don't know what "clicking an icon" means, then you need to take a basic Introduction to Computers course before proceeding.

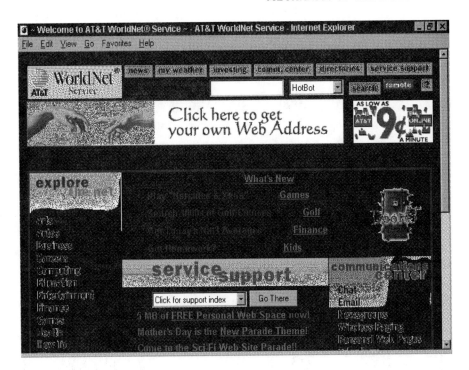

Clicking the Internet icon starts a program that uses your modem to call your ISP's computer and logs you onto it. The ISP's computer (server) is connected to the Internet all of the time. A server is any computer connected directly to the Internet.

After your connection is made, your computer launches a program called a browser. Netscape and Microsoft are the major publishers of browsers. An opening page comes up. Your opening page is set in your browser's setup feature and can be changed. Usually you start at the home page of your ISP, which is filled with ISP news, advertisement banners, and links to useful sites. On AT&T WorldNet the links might be news, weather, chat rooms, support services, etc.

You are now live on the World Wide Web (called WWW). Whenever you want to visit a site you can type in its location or address (called a URL, Universal Resource Locator) at the address line and press the enter key. Within seconds you are surfing the web. Addresses to visit can be found almost anywhere including (this book!), advertisements, magazine articles, TV commercials, books, or on web sites themselves.

If you move your cursor over the web site it sometimes turns into a hand with a finger pointing. The shape of the cursor changes when it is over a hyperlink. A hyperlink is an address of another page. Pushing the left mouse button loads the address and tells the computer to download that page. A hyperlink URL can be located on the original server or on any server in the world. Now you are really surfing!

WHAT DOES IT REALLY MEAN TO SURF THE NET?

When you enter a URL and press the enter key, you are actually making a request for a file. The file is a web page written in Hypertext Markup Language (called HTML). Browsers can convert HTML into a color web page. The simple request starts a series of actions that end up with you viewing the page requested.

Your server looks up the name of the site and translates it into the IP address (Internet Protocol, IP addresses are made up of all numbers). For example, the author's domain name is maintrainer.com, the IP address is 38.223.66.6 (Please note, I had to call my ISP to find out what my IP address was, since I never use it directly.) The list of IP addresses and domain names is kept up to date by the InterNIC. InterNIC is an agency that assigns domain names. After a domain name is approved it is issued to the DNS servers (Domain Name Servers are also called root servers) all over the world. The DNS servers are the core of the Internet. When we registered our domain name with Internic it only took 60 seconds for it to send the update command (with our IP address and domain name) to all the DNS servers in the world! Of course, that was on a Sunday.

Once your server has the address decoded into an IP number, it sends a request to that server for the file of the web page you requested. The request travels along the network wires from node to node. Each node has a router that looks at the address and routes the request to the next node that is nearer to the one you want. Your request for a web page might pass through ten or more nodes on the way to the server.

The server gets the request with the file name and locates the web page from its hard drive. All web pages are files on hard drives with specific file names and paths. The server then sends the file (web page) to its modem and router and starts the same process back to your server. The web page file gets passed from node to node until it reaches you.

When response is slow, it usually means that many messages are being sent at the same time. The routers are queuing the messages (putting the messages into a line) because of the volume. Videos, music and complex pictures absorb much of the capacity. Capacity is called bandwidth.

Web pages with lots of pictures, sounds or video clips take up bandwidth and may move slowly over the Internet. Many people set their browsers to turn off the pictures, video and sound, in order to speed the process. It's not as much fun, but it does work.

The page you requested appears on your screen. If you have a conventional hook-up from home, you may complain about how long it takes or how slow the Internet is running that day. There are many solutions to the speed problems being sold to the small office and home markets. Some solutions include using cable TV wires and special digital telephone lines (called ISDN).

WHO PAYS?

One of the reasons the Internet has grown so explosively is its low cost of use. The Internet is funded by the military, universities and organizations that setup servers. These servers are all linked together and are the nodes of the network. Each server pays for the high speed communications line to a central backbone that snakes around the world.

Individuals and small businesses that cannot afford the communications fees and hardware (in 1998 the costs were about $100-$500/month for a dedicated communications line plus $7500 or more for the server and communications equipment) can rent access to the Internet through service providers already mentioned. An ISP that allows you to keep your web page on its server is called a Web hosting service. Total costs for Web hosting range from about $50 for a simple site to $1000 per month for a very large and active site with direct sales. Many of the ISPs also allow a small, low volume personal web page at no additional charge (to your monthly subscription fee).

INTERNET ADDRESSES

Each address on the Internet is unique. The address has several parts and cannot have commas or spaces. Some sites include the path to the specific directory where the information is stored. In UNIX (the traditionally most popular operating system for the Internet) forward slashes are used to show the path to the directory (rather then backward slashes that are used on WINTEL PCs) where the web page is stored.

Parts of an Internet address

Protocol: The first part of the address is the protocol. Web pages start with http://. This tells the browser to use Hypertext Transfer Protocol. Other protocols could be ftp://, telnet://, and gopher://. Modern browsers can identify what is needed to use each protocol. You don't have to know anything about the differences between them to use them and enjoy the benefits.

Domain: This is the last part after the period. It puts you into a large category like .com for commercial, .edu for education, .gov for government.

Outside the United States, the domain might be a country such as .uk for Great Britain or .fr for France.

Organization or location: Each user is part of an organization. For example, commercial organizations try to get their name or initials such as omega.com for Omega Corporation or GE.com for General Electric. Some try to get a related slogan or memorable phrase for the organization like maintrainer for a company that trains maintenance professionals. In E-mail addresses the @ symbol (pronounced at) is the divider betwccn the user name "jdl" and the organization name "maintrainer". The organization name follows the @ symbol.

User name: This is is the unique name of the specific user within an organization such as "jdl". Each user is assigned a name within the organization and domain. A user name (like johnjones) could be duplicated in another organization but not in that organization. Many organizations use some variant of the first initial and the last name such as jlevitt for Joel Levitt. They might add a middle initial if the first one is taken.

The author's E-mail address:	jdl@maintrainer.com
	User name@Organization.Domain
The author's web page URL is:	http://www.maintrainer.com
	Protocal:Organization.domain

A tip on extended addresses (the ones with slashes after the domain like www.maintrainer.com/newsletters/april.html): If the address you are looking for gives you an error message and says the site is not found, try the following. Strip off the last part of the URL, and try again. Keep stripping off parts of the address until you reach the domain. Frequently a webmaster will change the name of one page (the one you are looking for, of course).

So if www.maintrainer.com/newsletters/april.html doesn't work try www.maintrainer.com/newsletters and if that doesn't work try www.maintrainer.com/. You may get lucky and pickup a reference to the new address of the page you are looking for.

HOW YOU REMEMBER THE PLACES THAT YOU VISIT

Modern browsers have two capabilities to help users remember where they have been. The one feature is usually called either bookmarks or favorites. When you find a site you like, you activate the function and the browser places the address and description in a structured list.

As you can see the bookmarks are arranged like the folder structure of Windows 95 ® or the files structure of Windows 3.1®. You create a

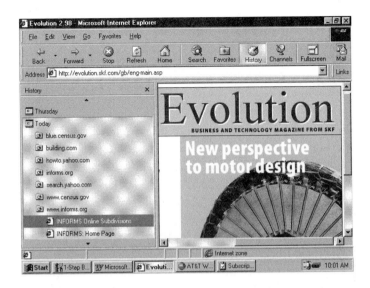

hierarchy with the big topics at the top and the sub-topics under them. The big (highest level topics might be work, play, kids, travel and shopping for a home computer. The work level might breakdown to engineering, vendors, consultants, competitors, etc. You can make as many levels as you want.

Browsers also keep track of all the sites they have visited for the last 30 days. Remember this if you are visiting sites that are innapproiate at work (or even at home, for that matter). Activate the history function through the menu bar at the top of the screen. Browsers allow you to change the number of days the history is kept.

Once activated the history file is displayed by day starting with the current day. Clicking the up triangle moves the history further into the past. Clicking the day opens that day's Internet session.

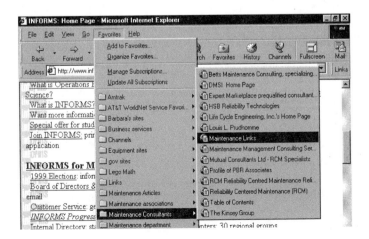

HOW DO YOU FIND THINGS AMONG THE 80,000,000 DIFFERENT WEB PAGES?

Soon after the WWW was developed, programmers developed search engine sites. Search sites (there are hundreds now) narrow the search down from millions of sites to a few hundred or less. A well-designed search can save hundreds of hours.

The search engines use different techniques to catalog the sites. Some search sites require you to submit a request to include your site in their directory. You give them the key words that you want your site indexed under. The first popular search site YAHOO is designed this way.

Other sites use programs called spiders to search all the sites on the web one at a time. The spider takes the text of the site and creates an index of all the words. When you look up a topic on a spider search site, it reviews the index it's created of all the search results.

Each search site is different. Now search sites are big business because of advertising. The tops of the screen of many search engines contain advertisement banners. The advantage of banners is that if you are interested in the product you can surf over to the advertised site with a single click. In the best cases the advertisement is directly related to the topic that you searched. On one of my searches on maintenance, a banner came up featuring books for sale on maintenance topics. The chapter on search engines will demonstrate the different sites.

FTP (FILE TRANSFER PROTOCOL)

FTP is an Internet tool that will allow you to transfer files from one computer to another computer. It allows you to transfer files between two computers faster than you could by using programs like Kermit, terminal, xmodem, etc.

FTP is a standard about how files transfers will be handled on the Internet. In the pre-Internet days if two people wanted to exchange files they had to agree to both a hardware and software protocol. The protocols include information about record markers, file markers, communications speeds, word length, type of transfer (there are several), file name conventions and several other issues.

FTP simplified file transfer by making those decisions for you. File transfer is as simple as clicking a few buttons. The most difficult thing about file transfer (this has caught me and almost everyone I know at least once)is finding the file on your hard disk after it is transferred.

Windows 95(r) has partially solved this problem (once you know where to look). In the C: drive on a single user system (on networks ask your system administrator for the drive location of this folder) look in the Program Files folder. Once in Program files click the Downloaded files folder.

FTP Hint: Following the belt and suspenders technique of computing always write the exact file name down before the transfer starts. If you have the file name, you can use the Find file function (right click the Start button, left click Find).

FTP has two main applications. The first is obtaining publicly accessible files from other systems on the Internet. Second, you can use FTP to transfer files between a desktop system connected to a network and other systems connected to the Internet both inside and outside your company. For example, FTP provides a fast way to transfer files from a computer in a University computing site to your computer (and vice versa).

Obtaining Publicly Accessible Files

There are many kinds of files available to users of the Internet, ranging from research data to archives of information to shareware for different computers. Often, people will refer to these files in postings made to a Usenet news group or in a mail message.

Anonymous FTP

Let's say you know the location of a file that you want to retrieve. In order to do so, you need to log in to the computer where the file is located. Many sites allow "anonymous ftp" logins; that is, you are allowed to log in to that system, even though you do not have an account on that system, and get files from that system. In short, anonymous FTP allows those with Internet access to login to remote computers for the purpose of transferring publicly accessible files.

The Internet Tutorial at the University of Delaware on FTP has an illustration of anonymous FTP. CAVEAT: Anonymous FTP sites are often available to a limited number of users at any one time. This can prove frustrating if you try to access popular FTP sites during rush hour (e.g., during the business day at that site).

Archie

If you want to search for files of information on a topic of interest to you, there are a variety of Internet tools available to you. One of the most widely used is Archie. Archie is a database of the names of files that are available to you via Anonymous FTP. There are several different sites that maintain copies of the Archie database. Netiquette suggests you access the site nearest you. Each site contains a copy of the same database of filenames.

Since Archie is a database of file names, you will see full "UNIX-style" pathnames for the files. That is, you will see the name of the computer that has the file, the name of the file, and the directory in which that file is stored.

For serious FTP usage consider acquiring FTP client software such as WS-FTP95 Professional (available from www.ipswitch.com for $37.50). This software allows you to search an FTP directory, pick out the files you want to download and manage the download task. The screens look like the Explorer file manager from Windows 95 and the operation is similar.

One advantage of the Internet is you can decide that you want to look at software like this FTP client system, you can go to the site and download a 30-day demo and go to work, all in 30 minutes.

Software that that has been a boon to the Internet in general and to FTP in particular is compaction software. Most compaction software is known by the trade name of one of the popular products called Zip (the program is called Winzip). In computer parlance you zip to compress a file and unzip to uncompress a file. Depending on the type of file zipping can reduce its size by 75%. This compaction reduces the download time. Some software downloads have taken me 5 hours with compaction!

Winzip utility is available on the Internet for download at http://www.winzip.com for $22, or by sending postal mail to P.O. Box 540, Mansfield, CT 06268-0540 USA.

Another important product to know about is called Adobe Acrobat. Acrobat creates files in .pdf format (Portable Document Format). Most manuals and reports are stored on the web in .pdf format because it allows the author's formatting, font size and color, and images be preserved across many platforms (different kinds of computers). The Acrobat viewer is free. It is called a plug-in to your browser. Almost every site that uses Acrobat also has a button to take you to the Adobe download site, for the free download.

Every few months a new format is introduced for audio, video, document, compression, or encryption. The only way you can stay abreast of the changes is to read one of the Internet magazines, follow the computer press or spend some time surfing. Whenever you reach a site you can't look at (or see, or listen to), download whatever viewer or plug-in is needed.

How to act when online, Netiquette Guidelines

In the past, the population of people on-line had grown up with the Internet, was technically minded, and understood the nature of the transport and the protocols. Today, the community of Internet users includes people who are new to the environment. These "newbies" are unfamiliar with the culture and don't need to know about transport and protocols. This section is designed to bring these new users into the Internet culture quickly.

I would like to thank Sally Hambridge of Intel Corporation for permission to reprint an edited version of her excellent memo on netiquette. She can be reached at Email: sallyh@ludwig.sc.intel.com

One-to-one Communication (electronic mail, talk)

One-to-one communication is when you are communicating with another person as if face-to-face. In general, rules of common courtesy for interaction with people should be in force for any situation. On the Internet, it's doubly important because body language and tone of voice must be inferred.

For E-mail: Unless you have your own Internet access through an Internet provider, be sure to check with your employer about ownership of electronic mail. Laws about the ownership of electronic mail vary from place to place. Your employer might own your mail!

Assume that mail on the Internet is not secure. Never put in an E-mail message anything you would not put on a postcard. Respect the copyright on material that you reproduce. Also assume that your message can easily be forwarded to anyone else without your knowledge.

If you are forwarding a message you've received, do not change the wording. If the message was a personal message to you and you are re-posting to a group, you should ask permission first. You may shorten the message and quote only relevant parts, but be sure you give proper attribution.

Never send chain letters via electronic mail. Chain letters are forbidden on the Internet. Notify your ISP or local system administrator whenever you receive one.

A good rule of thumb: Be conservative in what you send and liberal in what you receive. You should not send heated messages (we call these "flames") even if you are provoked. On the other hand, you shouldn't be surprised if you get flamed. It's prudent not to respond to flames.

In general, it's a good idea to check all your mail subjects before responding to a message. Sometimes a person who asks you for help (or clarification) will send another message which effectively says "Never Mind." Also make sure that any message you respond to was directed to you. You might be copied (cc:) rather than the primary recipient.

Make things easy for the recipient. Many mailers strip header information, which includes your return address. In order to ensure that people know who you are be sure to include a line or two at the end of your message with contact information. You can create this file ahead of time and add it to the end of your messages. (Some mailers do this automatically.) In Internet parlance, this is known as a ". SIG" or signature file. Your .SIG file takes the place of your business card. (Like business cards you can have more than one to apply in different circumstances.)

Be careful when addressing mail. There are addresses that may go to a group even though the address looks like it is just one person. Know to whom you are sending. Watch cc's when replying. Don't continue to include other people if the messages have become a two-way conversation.

Remember that people with whom you communicate are located across the globe. If you send a message to which you want an immediate response, the person receiving it might be at home asleep when it arrives. Give them a chance to wake up, come to work, and log in before assuming the mail didn't arrive or that they don't care.

Verify all addresses before initiating long or personal discourse. It's also a good practice to include the word "Long" in the subject header so the recipient knows the message will take time to read and respond to. Over a hundred lines is considered long.

Know whom to contact for help. Usually you will have resources close at hand. Check locally for people who can help you with software and system problems. Also, know whom to go to if you receive anything questionable or illegal. Most sites also have users called "Postmaster" or "Webmaster" who are knowledgeable site administrators; so you can send questions to this address to get help with mail.

Remember that the recipient is a human being whose culture, language, and humor may have different points of reference than your own. Remember that date formats, measurements, and idioms may not travel well. Be especially careful with sarcasm and humor.

Use mixed case.

UPPER CASE LOOKS AS IF YOU'RE SHOUTING.

Use symbols for emphasis.

That *is* what I meant. Use underscores for underlining.

War and Peace is my favorite book.

Use smileys (also called emoticons) to indicate tone of voice, but use them sparingly. The Internet users developed the smileys to connote emotional states or tone.:-<, :-) are examples of a smileys (look at them sideways). Don't assume that the inclusion of a smiley will make the recipient happy with what you say or wipe out an otherwise insulting comment.

Wait overnight to send emotional responses to messages. If you have really strong feelings about a subject, indicate it via FLAME ON/OFF enclosures. For example:

FLAME ON This type of argument is not worth the bandwidth it takes to send it. It's illogical and poorly reasoned. The rest of the world agrees with me.

FLAME OFF If you send encoded messages make sure the recipient can decode them. The same goes for enclosures and attachments. Make sure the recipient has the software to open them..

Be brief without being overly terse. When replying to a message, include enough original material to be understood, but no more. It is extremely bad form to simply reply to a message by including the entire previous message: edit out all the irrelevant material.

Limit line length to fewer than 65 characters and end a line with a carriage return.

Mail should have a subject heading which reflects the content of the message.

If you include a signature, keep it short. The rule of thumb is no longer than four lines. Remember that many people pay for connectivity by the minute. The longer your message is, the more they pay.

Just as E-mail may not be private, E-mail and news group postings are subject to forgery and spoofing with various degrees of delectability. Apply common sense reality checks before assuming a message is valid.

If you think the importance of a message justifies it, immediately reply briefly to an e-mail message to let the sender know you got it, even if you will send a longer reply later.

Reasonable expectations for conduct via e-mail depend on your relationship to a person and the context of the communication. Norms learned in a particular e-mail environment may not apply in general to your e-mail communication with people across the Internet. Be careful with slang or local acronyms.

The cost of delivering an e-mail message is, on the average, paid about equally by the sender and the recipient (or their organizations). This is unlike other media such as physical mail, telephone, TV, or radio. Sending someone mail may also cost them in other specific ways like network bandwidth, disk space or CPU usage. This is a fundamental economic reason why unsolicited e-mail advertising is unwelcome (and is forbidden in many contexts).

Know how large a message you are sending. Including large files such as Postscript files or programs may make your message so large that it cannot be delivered or, at least, consumes excessive resources. A good rule of thumb would be not to send a file larger than 50 kilobytes. Consider file transfer as an alternative, or cut the file into smaller chunks and sending each as a separate message.

Don't send large amounts of unsolicited information to people.

For talk, chat: Talk is a set of protocols that allow two people to have an interactive dialogue via computer.

Use mixed case and proper punctuation, as though you were typing a letter or sending mail. Don't run off the end of a line and simply let the terminal wrap; use a Carriage Return (CR) at the end of the line. Also, don't assume your screen size is the same as everyone else's. A good rule of thumb is to write out no more than 70 characters, and no more than 12 lines (since you're using a split screen).

Leave some margin; don't write to the edge of the screen.

Use two CRs to indicate that you are done and the other person may start typing. (blank line).

Always say goodbye, or some other farewell, and wait to see a farewell from the other person before killing the session. This is especially important when you are communicating with someone a long way away. Remember that your communication relies on both bandwidth (the size of the pipe) and latency (the speed of light).

Remember that talk is an interruption to the other person. Only use as appropriate. And never talk to strangers.

The reasons for not getting a reply are many. Don't assume that everything is working correctly. Not all versions of talk are compatible. If left on its own, talk re-rings the recipient. Let it ring one or two times, then kill it.

Talk shows your typing ability. If you type slowly and make mistakes when typing, it is often not worth the time of trying to correct, as the other person can usually see what you meant.

Be careful if you have more than one talk session going!

One-to-Many Communication (Mailing Lists, NetNews)

Any time you engage in one-to-many communication, all the rules for E-mail also apply. After all, communicating with many people via one mail message or post is quite analogous to communicating with one person, with the exception of possibly offending a great many more people than in one-to-one communication. Therefore, it's quite important to know as much as you can about the audience of your message.

Read both mailing lists and newsgroups for one to two months before you post anything (this is called net lurking). This helps you to get an understanding of the culture of the group.

Do not blame the system administrator for the behavior of the system users.

Consider that a large audience will see your posts. That may include your present or your next boss. Take care with what you write. Remember too, that mailing lists and newsgroups are frequently archived. Your words may be stored for a very long time in a place to which many people have access.

Assume that individuals speak for themselves and what they say does not represent their organization (unless stated explicitly).

Messages and articles should be brief and to the point. Don't wander off-topic, don't ramble, and don't send mail or post messages solely to point out other people's errors in typing or spelling. These, more than any other behavior, mark you as an immature beginner.

Subject lines should follow the conventions of the group.

Forgeries and spoofing are not approved behavior.

Advertising is welcomed on some lists and newsgroups, abhorred on others! This is another example of knowing your audience before you post. Unsolicited advertising that is completely off-topic will most certainly guarantee that you get a lot of hate mail.

If you are sending a reply to a message or a posting be sure you summarize the original at the top of the message, or include just enough text of the original to give a context. This will make sure readers understand when they start to read your response. Since NetNews, especially, is proliferated by distributing the postings from one host to another, it is possible to see a response to a message before seeing the original. Giving context helps everyone. But do not include the entire original!

Again, be sure to have a signature that you attach to your message. This will guarantee that any peculiarities of mailers or newsreaders that strip header information will not delete the only reference in the message of how people may reach you.

Be careful when you reply to messages or postings. Frequently replies are sent back to the address that originated the post, which in many cases is the address of a list or group! You may accidentally send a personal response to a great many people, embarrassing all involved. It's best to type in the address instead of relying on "reply."

Delivery receipts are invasive when sent to mailing lists, and some people consider delivery receipts an invasion of privacy. In short, do not use them.

If you find a personal message has gone to a list or group, send an apology to the person and to the group. If you should find yourself in a disagreement with one person, make your responses to each other via direct E-mail rather than continue to send messages to the list or the group. If you are debating a point on which the group might have some interest, you may summarize for them later.

Don't get involved in flame wars. Neither post nor respond to incendiary material.

Avoid sending messages or posting articles that are no more than gratuitous replies to replies.

Be careful with monospace diagrams and pictures. These will display differently on different systems, and with different mailers on the same system.

There are newsgroups and mailing lists that discuss topics of wide varieties of interests. These represent a diversity of lifestyles, religions, and cultures. Posting articles or sending messages to a group whose point of view is offensive to you simply to tell them they are offensive is not acceptable.

Sexually and racially harassing messages may also have legal implications. There is software available to filter items you find objectionable.

Special Guidelines for Mailing List

A mailing list is related to a newsgroup. The difference is that a mailing list is based on E-mail and the newsgroup is based on news protocols. You post messages on a mailing list by sending them to a special address. All mail sent to that address is reflected to all members of the mailing list. Some lists are moderated where someone reads all incoming mail and will not print postings that are not related to the topic.

There are several ways to find information about what mailing lists exist on the Internet and how to join them. There are a set of files posted periodically to news.answers that list the Internet mailing lists and how to subscribe to them. This is an invaluable resource for finding lists on any topic.

Send subscribe and unsubscribe messages to the appropriate address. Although some mailing list software is smart enough to catch these, not all can ferret these out. It is your responsibility to learn how the lists work, and to send the correct mail to the correct place. Although many mailing lists adhere to the convention of having a "request" alias for sending subscribe and unsubscribe messages, not all do. Be sure you know the conventions used by the lists to which you subscribe.

Save the subscription messages for any lists you join. These usually tell you how to unsubscribe as well. In general, it's not possible to retrieve messages once you have sent them. Even your system administrator will not be able to get a message back once you have sent it. This means you must make sure you really want the message to go as you have written it.

The auto-reply feature of many mailers is useful for in-house communication, but quite annoying when sent to entire mailing lists. Examine "Reply-To" addresses when replying to messages from lists. Most auto-replies will go to all members of the list.

Don't send large files to mailing lists when Web page URLs or pointers to FTP versions will do. If you want to send it as multiple files, be sure to follow the culture of the group. If you don't know what that is, ask.

Consider unsubscribing or setting a "nomail" option (when it's available) when you cannot check your mail for an extended period.

When sending a message to more than one mailing list, especially if the lists are closely related, apologize for cross posting. If you ask a question, be sure to post a summary. When doing so, truly summarize rather than send a cumulation of the messages you receive.

Some mailing lists are private. Do not send mail to these lists uninvited. Do not report mail from these lists to a wider audience.

If you are caught in an argument, keep the discussion focused on issues rather than the personalities involved.

SPECIAL GUIDELINES FOR NET NEWS

Net News is a globally distributed system that allows people to communicate on topics of specific interest. It is divided into hierarchies, with the major divisions being:

sci - science-related discussions
comp - computer-related discussions
news - for discussions which center around NetNews itself
rec - recreational activities
soc - social issues
talk - long-winded never-ending discussions
biz - business related postings
alt - the alternate hierarchy.

Alt is so named because creating an alt group does not go through the same process as creating a group in the other parts of the hierarchy. There are also regional hierarchies, hierarchies that are widely distributed, such as Bionet. Your place of business may have its own groups as well. Recently, a humanities hierarchy was added and, as time goes on, it's likely more will be added

In NetNews parlance, "posting" refers to posting a new article to a group, or responding to a post someone else has posted. "cross-posting" refers to posting a message to more than one group. If you introduce cross-posting to a group, or if you direct "Follow-up To:" in the header of your posting, warn readers! Readers will usually assume that the message was posted to a specific group and that follow-ups will go to that group. Headers change this behavior.

Read all of a discussion in progress (we call this a thread) before posting replies. Avoid posting "Me Too" messages, where content is limited to agreement with previous posts. Content of a follow-up post should exceed quoted content.

Send E-mail when an answer to a question is for one person only. Remember that News has global distribution and the whole world probably is NOT interested in a personal response. However, don't hesitate to post when something will be of general interest to the Newsgroup participants.

Check the "Distribution" section of the header, but don't depend on it. Due to the complex method by which News is delivered, Distribution headers are unreliable. But, if you are posting something that will be of interest to a limited number or readers, use a distribution line that attempts to limit the distribution of your article to those people. For example, set the Distribution to be "nj" if you are posting an article that will be of interest only to New Jersey readers. If you feel an article will be of interest to more than one newsgroup, be sure to CROSSPOST the article rather than individually post it to those groups. In general, probably only five-to-six groups will have similar enough interests to warrant this.

Consider using Reference sources (computer manuals, newspapers, help files) before posting a question. Asking a newsgroup where answers are readily available elsewhere generates grumpy "RTFM" (read the fine manual - although a more vulgar meaning of the word beginning with "f" is usually implied) messages.

Although there are Newsgroups that welcome advertising, in general it is considered nothing less than criminal to advertise off-topic products. Sending an advertisement to each and every group will pretty much guarantee your loss of connectivity.

If you discover an error in your post, cancel it as soon as possible.

DO NOT attempt to cancel any articles but your own. Contact your administrator if you don't know how to cancel your post, or if some other post, such as a chain letter, needs canceling.

If you've posted something and don't see it immediately, don't assume it's failed and re-post it.

Some groups permit (and some welcome) posts which in other circumstances would be considered to be in questionable taste. Still, there is no guarantee that all people reading the group will appreciate the material as much as you do. Use the Rotate utility (which rotates all the characters in your post by 13 positions in the alphabet) to avoid giving offense. The Rot13 utility for Unix is an example.

In groups, which discuss movies or books, it is considered essential to mark posts that disclose significant content as "Spoilers". Put this word in your subject line. You may add blank lines to the beginning of your post to keep content out of sight, or you may rotate it.

Forging of news articles is generally censured. You can protect yourself from forgeries by using software that generates a manipulation detection fingerprint, such as PGP (in the US).

Postings via anonymous servers are accepted in some newsgroups and disliked in others. Material that is inappropriate when posted under one's own name is still inappropriate when posted anonymously.

Expect a slight delay in seeing your post when posting to a moderated group. The moderator may change your subject line to have your post conform to a particular thread.

Don't get involved in flame wars. Neither post nor respond to incendiary material.

INFORMATION SERVICES (GOPHER, WAIS, WWW, FTP, TELNET)

In recent Internet history, the Net has exploded with new and varied information services. Gopher, WAIS, World Wide Web (WWW), Multi-User Dimensions (MUDs) Multi-User Dimensions which are Object Oriented (MOOs) are a few of these new areas. Although the ability to find information is continuing to expand rapidly, "Caveat Emptor" remains constant.

Remember that all these services belong to someone else. The people who pay the bills get to make the rules governing usage. Information may be free - or it may not be! Be sure you check.

If you have problems with any form of information service, start your problem solving by checking locally: Check file configurations, software setup, network connections, etc. Do this before assuming the problem is at the provider's end or is the provider's fault.

Although there are naming conventions for file types used, don't depend on these file-naming conventions to be enforced. For example, a ".doc" file is not always a Word file.

Information services also use conventions, such as www.xyz.com. While it is useful to know these conventions, again, don't necessarily rely on them.

Know how file names work on your own system. Be aware of conventions used for providing information during sessions. FTP sites usually have files named README in a top-level directory which have information about the files available. But don't assume that these files are necessarily up-to-date or accurate.

Do NOT assume that ANY information you find is up-to-date or accurate. Remember that new technologies allow just about anyone to be a publisher, but not all people have discovered the responsibilities that accompany publishing.

Remember that unless you are sure that security and authentication technology is in use, any information you submit to a system is being transmitted over the Internet "in the clear," with no protection from "sniffers" or forgers.

Since the Internet spans the globe, remember that information services might reflect culture and life-styles markedly different from your own community. Materials you find offensive may originate in a geography which finds them acceptable. Keep an open mind.

When wanting information from a popular server, be sure to use a mirror server that's close if a list is provided. Do not use someone else's FTP site to deposit materials you wish other people to pick up. This is called dumping and is not generally acceptable behavior.

When you have trouble with a site and ask for help, be sure to provide as much information as possible in order to help debug the problem. When bringing up your own information service, such as a homepage, be sure to check with your local system administrator to find what the local guidelines are in affect.

Consider spreading out the system load on popular sites by avoiding rush hour and, logging in during off-peak times.

REAL TIME INTERACTIVE SERVICES GUIDELINES (MUDS MOOS IRC)

As in other environments, it is wise to listen first to get to know the culture of the group.

It's not necessary to greet everyone on a channel or room personally. Usually one "Hello" or the equivalent is enough. Using the automation features of your client to greet people is not acceptable behavior.

Warn the participants if you intend to ship large quantities of information. If all consent to receiving it, you may send, but sending unwanted information without a warning is considered bad form, just as it is in E-mail.

Don't assume that people who you don't know will want to talk to you. If you feel compelled to send private messages to people you don't know, then be willing to accept gracefully the fact that they might be busy or simply not want to chat with you.

Respect the guidelines of the group. Look for introductory materials for the group. These may be on a related ftp site.

Don't badger other users for personal information such as sex, age, or location. After you have built an acquaintance with another user, these questions may be more appropriate, but many people hesitate to give this information to people with whom they are not familiar.

If a user is using a nickname alias or pseudonym, respect that user's desire for anonymity. Even if you and that person are close friends, it is more courteous to use his nickname. Do not use that person's real name online without permission.

Additional material for this section was adapted from *Demystifying the Internet* at The University of Delaware. While it was written 1995, the information is even more valid today because of the numbers of newbies.

WEB HOSTING SERVICE

ISP can also provide space for your web page. This function is called web hosting. The host can provide extensive support services and design services for your site. Our web host (http://plhnet.com) has been instrumental in the sucess of the maintrainer.com site. The president, Brian Centonze, has frequent contributions, as any good web host should. There are many powerful tools to ease the design of web pages. Help from a professional is important, especially when you plan to use the site for business. If you want to design a web page consult with an expert like Brian.

*T*OUR OF THE *W*ORLD *W*IDE *W*EB

Any tour starts at a transportation center like an airport, train station or port. On the Internet the transportation centers are called supersites. These sites are filled with hyperlinks to related sites, similar to a packaged tour. You can also design a tour yourself by typing in individual URLs (site address-Universal Resource Locators) that you want to visit.

One of the fears of management (a well-founded one) is that maintenance professionals will go on the web and get lost for hours at a time. This can easily happen. To prevent this happening, I suggest that you both schedule your sessions and limit the surfing time. This Internet tour can take as little as ten minutes or as long as ten or more hours if you follow many leads (take side trips).

You might want to make a journal of your trip. Your browser contains a history file of all sites recently visited, which you can view. This history file is generally overwritten on a rolling 20-to-30 day basis (the number of days is set in your browser.). Your browser also has the capability to permanently record important addresses of your trip. Whenever you are on an interesting page activate your bookmark feature and the site's URL will be recorded.

The great thing about an Internet tour is you can get off the tour at any time to explore a site in depth. You can then go on your own and follow the site's links wherever they take you. I encourage you to stay as long as you like and look at any pages of interest. Jump back on the tour whenever you want by using the bookmark to return and jumping back to http://www.main-trainer.com. Alternatively you can rejoin to the tour by pushing the back button until you return. A third way is to manually type the URL that you want to visit.

When on a trip people have to satisfy life's needs in a foreign place. Similarly when we visit sites on the Internet, we also have to see that our needs are satisfied with what the site has to offer. Each site has to provide different capabilities that answer questions we have. Thus a virtual bookstore has to have tons of books, great prices or some great specialty for us to be satisfied with the visit.

As you visit each type of site, ask yourself a simple question, What would I be looking for if I were to visit this site in my work environment? We will look more deeply at each class of site in subsequent chapters and address that question.

PLACES WE WILL VISIT ON THIS TOUR:

Maintenance consultant and supersite: Springfield Resources (http://www.maintrainer.com)
This is the author's site. It lists most of the sites in this book as hyperlinks. If you find a site you want to visit, just click on it. This site is organized the same way that the book is organized.

> If you find that any links have expired and give you an error, please send me an E-mail message: jdl@maintrainer.com so I can track down the new URL. Also, if you find a great site that I missed, please send me a message so I can include the URL.

Find the supersite section then find Vibration Institute of Canada for the first stop of the tour.

Maintenance supersite: Vibration Institute of Canada (http://www.vibrate.net)
This site has dozens of links to other sites, newsgroups, and associations. Maintenance professionals will be very interested in many of the organizations listed. Associations are generally good sources for leads to interesting sites in a particular industry.

Research site (http://cdr.stanford.edu/html/WWW-ME/home.html)
Stanford is a major university for research. Its virtual Mechanical Engineering Library is an excellent resource for research. This is a Supersite that lists resources in mechanical engineering throughout the university and outside organizations. Stanford also has a research site for predictive maintenance with dozens of links to other resources.

A research site of this type is like a magical card catalog. You review the entries (cards). When you find one that you want, instead of walking to the library stacks, you just push the button and get the resource brought to you.

Maintenance magazine: Plant Services http://www.plantservices.com

The maintenance magazines are also excellent supersites. Add an archive of articles, chat forums, advertisements and you have a potent blend. The magazine Plant Services has a useful service; they send you E-mail when the site has changed including the table of contents of the new issue.

Maintenance department web page http://www.ncboard.edu.on.ca/plant.htm

Many maintenance departments proudly promote their activities via a web page. Some offer service requests via the web. Most of the maintenance departments that I have found are from universities (where every department has a Web presence).

Equipment sites

We will visit a sample of the OEM equipment, component manufacturers, distributors and equipment dealers sites. The powerful advantage of the Internet comes from is that it provides you instant communications with this group. I recommend that someone in your maintenance or purchasing department make a list of the URLs of the organizations with which you do business.

General Pump (http://www.generalpump.com.)

General Pump of Mendota Heights, Minnesota, has developed a service on the Internet they call "Datalink." This site is up 24 hours a day, year around. It features downloadable engineering drawings, bills of materials for their pumps, new product information, discussion groups, training capabilities, and the ability to place electronic orders. Look closely because this is the future of the Internet.

Omega (http://www.omega.com)

The most user-friendly company in the transducer field is Omega. In fact, they are my favorite transducer company. Their colorful and useful catalogs have graced my shelf for almost two decades. I imagine that the printing costs for the catalog are quite high each year. When people start using the Web site they never want to go back to catalogs again. Companies with large, expensive catalogs like Omega or AMP can lower costs, speed dissemination of new information and improve customer services at the same time.

Government site of interest to maintenance: Department of Energy http://www.doe.gov

The government is the biggest user of the Internet. It has thousands of homepages for all agencies, departments and bureaus. The DOE has resources of use to the maintenance field in the area of energy conservation, research, and grants for energy reduction projects.

Industrial distributor: W.W. Grainger http://www.grainger.com

The 3 1/2" Grainger catalog has also been a mainstay reference book for whatever I needed since I entered the manufacturing world. Even as a consultant, I still order from it. W. W. Grainger is one of the largest industrial distributors in the world. It has had a proactive approach toward technology. The company's CD-ROM, which has been available for years, includes a proprietary interface to allow access to the Grainger order entry system. It is no surprise that Grainger has a pretty complete web site. Imagine the possibility of direct computer to computer linkage through the Internet with Grainger. By using these linkages, the acquisition cost of maintenance parts and service items has plummeted.

New and used machinery sales http://www.cayceco.com/index.html/

You can buy almost anything on the Internet. The used machinery companies have staked out a section of the web for their offerings. Their pages show typical layouts with master equipment lists, prices, equipment ages and, if you are lucky, a photograph.

Search engine: Yahoo! (http://www.yahoo.com)

Yahoo! was first. In a classic Internet story, some college kids wrote the program and created the search site. Their idea was almost an instant success. Yahoo! is one of the most popular search sites. Yahoo uses human knowledge engineers to create the indexes. Advertising banner revenue funds Yahoo! and most of the other search engines.

Job posting: http://www.columbia.edu/cu/hr/jobs/970713.html Columbia University

One of the most interesting uses of the Internet is job hunting. Sites now list job openings available. In the example, we visit a job posting from Columbia University for a Plant Engineer. It was advertised in early 1998.

Beware

One of the problems of the Internet is that it is not often easy to check the veracity of the information, articles, opinions and miscellaneous postings that you might find. There is no pride of accuracy that so often characterizes the print media. On the Internet, let the researcher beware. One partial way around this is to stick to major known sites. But even the most prestigious university sites accumulate false information. They are constantly on guard but some files, downloads and articles have been dumped onto their servers and are bogus!

When you strike out on your own, you will find that the URLs of sites change; some sites are taken off the server entirely. You will get a 404 error (site not found). This is common with individual specific pages. In fact, when you tour the sites recommended here, some might be missing! That means that between the time that we last checked the site (just before publication) and the time you read this book, the URL has changed. Many servers and modern browsers have URL forwarding that automatically moves you to the new URL.

One trick (mentioned in the last chapter) is address stripping. Starting from the right, strip off a section of the URL (to a period or slash) and keep retrying. Keep this up until you reach the domain e.g. com, net, org or edu. This technique moves you up the directory tree where it is more likely that you will find something.

ALL ABOARD, THE TOUR IS STARTING.

Consultant/ Supersite: Springfield Resources

Type http://www.maintrainer.com into your browser when you are live on the Internet. Pull down the bookmark or favorites menu and record this URL. This will make it easy for you to come back if you get lost. Click the button called Maintenance Management Resources. I've designed the site so you can directly access the sites in this book (the ones that allow a link). When you visit, send me a message by E-mail by clicking on the appropriate box. If, in your travels through cyberspace, you find any sites that you think I must include, for the next version, include the URL in the E-mail.

Maintenance supersite: Vibration Institute of Canada (http://www.vibrate.net)

From the transportation center on maintrainer.com click the Vibration Institute button (located in the supersite section). Once at the site notice the underlined items. Each of the underlined items on the left side is a hyperlink to another page on the Institute server. Each of these pages brings up a list of sites to visit. These lists are hyperlinks that point to pages on servers across Canada and the world.

Click a few and jump around on this server. Go out and visit any of the resources listed. To return to this site hit the back key until you come back, you can also pull down your bookmarks and click the Springfield Resources hyperlink, go to the transportation center and click the Vibration Institute hyperlink. Notice that this site has a registry that will notify you by E-mail when the site is changed. Registering makes your name available for related marketing (the same as magazines).

In the beginning of the Internet, people and organizations maintained sites for the good of the Internet citizens. For the first few years there was no profit motive. In fact, profit and commerce were banned from polite Internet

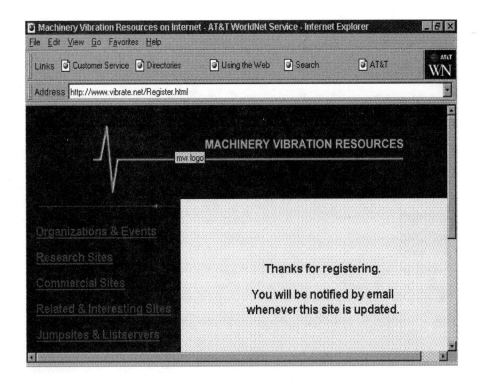

society. Sites like the Vibration Institute more in the original model, public services to the vibration community.

University library: Stanford
(http://cdr.stanford.edu/html/WWW-ME/home.html)

Research sites are at the core of the Internet. However, there are few pure maintenance research sites to visit at this time. One of the reasons is the dearth of serious university level research in the field. Only a few schools even offer courses in any aspect of maintenance. As time goes on, this problem will be solved, but for now we have to piggyback off of engineering, manufacturing and architecture libraries.

Maintenance magazine: Plant Services (http://www.plantservices.com)

Plant Services has several capabilities of interest to maintenance professionals. If you are like me, it is hard to get time to read the maintenance magazines. Yet I occasionally have to research a topic. Unless I have looked through the magazines and filed the articles, I'm at a loss for up-to-date research. One capability is an archive. You can look up old Plant Services' magazines without keeping the magazines on a pile on the desk or your file drawers full by clicking in the Past Issues area.

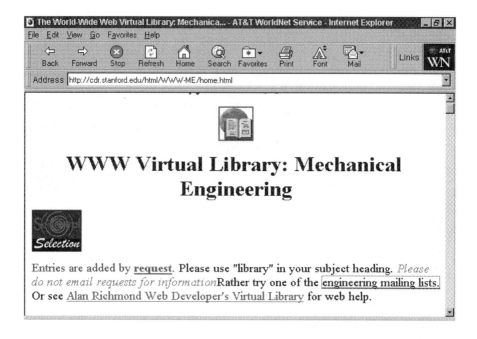

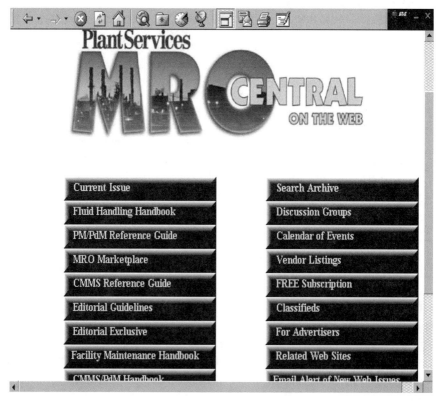

Some of the web magazines like Plant Services maintain newsgroups (Internet based discussions- see chapter 9 for details). These are moderated discussions, interviews and forums on maintenance related topics. If you have a particularly sticky technical problem, help might be a click away.

Another use of magazines is the information they provide through their annual buyer's guides. Many maintenance departments keep these guides so that they can quickly generate a list of potential bidders for a project. The new Internet based MRO market satisfies this need without the shelf space or the time and effort of filing, tracking and updating.

Maintenance people have strong opinions. A powerful part of your Internet experience is the ability to easily send your opinions to editors, authors, and other opinion leaders. Push a few buttons and you're on a blank page. The computer is obediently waiting for your pearls of wisdom. Sending your communication just takes pushing another button. No addressing, envelopes or stamps needed.

Spend a few minutes visiting the different parts of the site. You will have to sign a guest register to enter certain portions of the site. Again, remember that register listing is made available to organizations that want to reach you and your firm.

Maintenance department web page: Kawartha Pine Ridge District School Board (http://www.ncboard.edu.on.ca/plant.htm)

Web sites can serve many useful purposes for maintenance departments. In this case, the site is an introduction to the school district's maintenance department. Any parent, new teacher or interested party can benefit from this capsule description. Some additional things that would be useful would be an organizational chart, a communication channel (usually an E-mail link), and ultimately a way for teachers and administrators to directly request services through the site.

Three stages of Internet site (Kawartha Pine Ridge District School Board is at stage one)

1. The first incarnation of the web site is 'here we are.'
2. The next incarnation is always starting to use the media to have a dialog with the visitor (push this button to send a message to your friendly maintenance depart ment).
3. The third and subsequent incarnations use the power of the media to change the way business is conducted - (E Work Orders, E-commerce, communication with staff, customer service, etc.).

PLANT OPERATIONS MAINTENANCE AND CAPITAL

Plant Operations, Maintenance and Capital Services at the Cobourg Office of Kawartha Pine Ridge District School Board includes system-wide responsibility for custodial services, building maintenance and grounds keeping.

The Plant Department maintains approximately 2,460,000 square feet (228,542 square metres) of buildings made up of 53 Elementary Schools, 8 Secondary Schools, 4 Centres for Individual Studies, 3 Outdoor Education Centres, 1 Pioneer School, 1 Community Training and Development Centre, 1 Administration Building, 1 Maintenance Building and 1 Warehouse facility, plus 177 portable classrooms with an area of approximately 136,00 square feet (12,635 square metres).

Our Staff is comprised of 174 full time equivalent (FTE) Custodians, 17 FTE Maintenance personnel, 12 Supervisory staff and 2 Clerical staff, representing salaries and benefits of approximately $8,000,000.

In September of 1996 the Board opened one new school (Dr. Ross Tilley Public School) in Bowmanville and one replacement school (Newcastle Public School) in the Village of Newcastle.

Our school population continues to grow resulting in a new school planned for construction during the fall of 1997 and spring of 1998 with an anticipated opening of September 1998. The new school will be named Lydia Trull Public School and will be located on Avondale Drive in Courtice south of highway #2, west of Courtice Road. Lydia Trull Public School is being designed to accommodate grades JK to 8 with a Ministry Rated Capacity of 564 pupil places.

Over the past decade our "Footprint" has increased by 17% or 360,000 square feet (33,445 square metres) to accommodate increased enrolment and program requirements.

The total budget for the Plant Operations department for 1997 was approximately $11,188,000 which included custodial salaries and benefits, utilities, grass cutting, snow removal, waste removal and custodial supplies and equipment.

The total budget for Plant Maintenance and Capital Maintenance for 1997 was approximately $6,445,000 which included maintenance staff salaries and benefits, major and minor maintenance of the Board's facilities, vehicle maintenance, portable relocations and supplies.

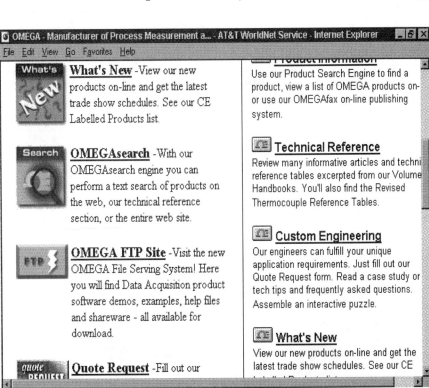

Component/sensor manufacturer: Omega (http://www.omega.com)

Omega has all of the attributes of a useful site for maintenance professionals. It has a logical organization. Most major questions asked by maintenance are answered on the home page. You can get answers, communicate to staff, search the site, get a map of the site, receive technical help, and visit Omega sites and affiliates around the world. You can even get a pithy quote for the day without changing channels.

Original Equipment Manufacturer: General Pump
(http://www.generalpump.com)

General Pump is one of the innovators in the use of the World Wide Web to provide customer service. They have made a commitment to providing a useful web presence. One day all manufacturers will fill their web sites with technical, maintenance, engineering as well as sales data. Many organizations also use their Web sites to advertise career opportunities. This is an interesting trend for maintenance Web surfers who are looking for a change of job or career.

Industrial Distributor: W.W. Grainger http://www.grainger.com

Industrial distributors like Graingers and McMaster-Carr have been a motivating force behind the computerization of maintenance. They have raised the bar for the entrance of new competitors. Their good Web business practices have shown the way for the smaller companies that follow.

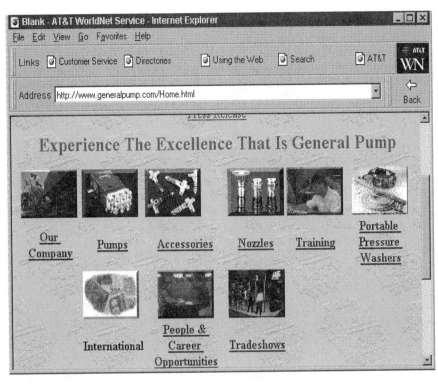

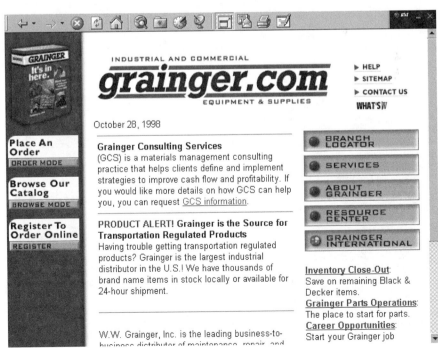

The Cayce Company has a wide variety of used machines at our warehouse. In addition to the machines in house, we always have access to many types of machines in the field that are also available. Please give us a call if you are looking for something and do not see it below.

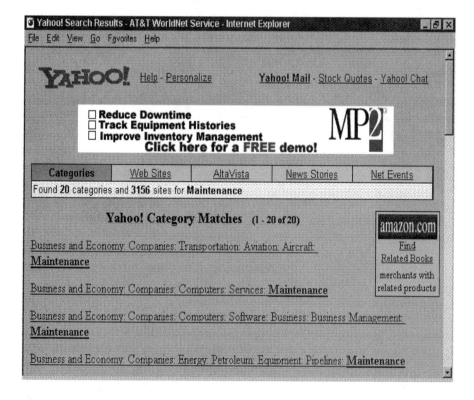

New and Used Equipment: Cayce Company
http://www.cayceco.com/index.html
    Buying and selling equipment and parts is already big business on the Internet. The Internet is particularly useful if the equipment is expensive or obscure. Expensive equipment justifies use of the Internet because saving a few tens of thousands of dollars is clearly worth the effort. Obscure equipment searching is justifiable because the users might be spread over the globe and the companies that supply the equipment might also be anywhere. The search can be sped up and the number of people that you can contact per hour is greater.

Search site: Yahoo! http://www.yahoo.com
    Yahoo was the WWW's first search site. It is very well organized and one of the most popular destinations on the World Wide Web.
    The first screen is the search screen. Type in the key words of the topic that you want searched. The second screen is the result screen. Yahoo found thousands of sites keyed to maintenance. Note the two advertisements on this screen. MP2 is a leading computer system for maintenance. Amazon.com is a giant virtual bookstore. If you click the Amazon.com button, a list of maintenance books, available for sale, will be shown.

Job posting: Columbia University
    Thousands of jobs are advertised on the Internet. As you can see from the example at Columbia University in New York, much more information can be fit into a web site then in a newspaper advertisement. This was available in 1998, type in up to /jobs/ if you can't find this specific page at http://www.columbia.edu/cu/hr/jobs/970713.html

Job Description

Manager
Reference #:970713
Date:09/15/1997
Grade:15
Department:Facilities Management
Salary:Commensurate with experience

Description: Reporting to the Director of Physical Plant, the incumbent is responsible for development and management of budget, material and staff resources for all Plant Engineering; oversees the day to day operation of Physical Plant Engineering group, which provides engineering support for plant capital, operating and maintenance projects; monitors work load of Plant Engineers including the development of project management approaches, designs, planning and implementation; monitors and assesses budget performance on projects and reports performance on a regular basis to the Director; manages work load input to the Plant Engineers to assure timely, quality response to University requests for service; establishes standards for new construction including specifications and details; continually assesses said standards and implements upgrades as required; coordinates with Operations and maintenance departments to develop and implement preventive maintenance programs; analyzes feedback from the system and adjusts programs; schedules and input for manpower requirements accordingly; works closely with other managers in the Division to insure a well-coordinated effort, prevent duplication of work, and insure the most effective and efficient use of resources; provides input into the development of the University's Capital Budget, and reviews impact of Capital Budget on programs; provides input into the University's Energy Conservation program; assesses equipment and building energy consumption, reviews operating procedures, develops and implements projects which will improve energy conservation; assesses technical and management expertise of personnel and assists in the development of appropriate training programs; provides input into safety programs to ensure a safe workplace and safety-conscious staff; reviews all staff performance evaluations and disciplinary communications; performs other related duties as assigned. MINORITIES ARE ESPECIALLY ENCOURAGED TO APPLY. Qualifications: Bachelor's degree in Mechanical or Electrical Engineering or related field required. Professional Engineer registration in the State of New York necessary. A minimum of seven years' related experience to include the design and troubleshooting of building systems required. Knowledge of NYC building codes, NFPA and NEC codes a must. Engineering experience from design through Project Management necessary. Experience managing in-house engineers and outside consultants required. Knowledge of Central Utility Plant Operations a must. Strong verbal and written communication skills necessary. Computer literacy and knowledge of Auto Cad desired.

This is a Morningside Campus-Officer- position. Qualified applicants should SEND A RESUME to Employment Office, Interchurch Center, 475 Riverside Drive, Room 1901, New York, NY 10115.

SEARCH ENGINES AND SUPERSITES

The most fascinating outgrowth of the World Wide Web is the search engine site. Many people would argue (I think accurately) that the WWW would not have grown as fast as it did if the search engine had not been developed. Search engines are WWW sites that can search the Internet for any topic. There are seven popular general-purpose search sites that are used for 90% or more of all the searches. In addition to these seven, there are about 300 smaller sites that provide the remainder of the searches.

Some of the smaller ones are extremely narrow in their searches. Examples include Findlaw for legal searches *http://www.findlaw.com* or *http://www.mapquest.com* for map searches. These are superior sources of information if you are looking for legal opinions or a map of an obscure area.

TWO TYPES OF SEARCH ENGINES: DIRECTORIES AND SPIDERS

It is important to know how the different engines work because some searches work better on some engines. The directory and the spider are two major types of search engines. The directory accepts requests from the owner of the web site and adds the site to their directory. The spider sends a program over the Internet that searches page by page and builds its index without contact from the webmaster of the site (although the spiders accept submissions from webmasters interested in speeding up the indexing process).

The first directory site, and the one that made the whole field popular by its success was Yahoo! The designers, Jerry Yang and David Filo, were graduate students in Electrical Engineering at Stanford. According to *Yahoo! Unplugged* (an IDG book), Yang and Filo got interested in the WWW while avoiding work on their dissertations.

For whatever reason, they started to spend hours each day searching and cataloging the web. They designed their own software to locate and index all the sites they were cataloging. They named the search site Yahoo!

The story goes that "Yahoo" might mean Yet Another Hierarchical Officious Oracle.

After a year on the web, Yang and Filo realized they had something because they were getting thousands of visits (hits) a day. Marc Andreessen, the designer of Mosaic (which became Netscape), saw the usefulness of the service and lent his support. Soon after, Randy Adams saw a future in advertising on the Yahoo! site and introduced the duo to Sequoia Financial, a major venture capitalist in Silicon Valley. Yahoo! became a business based on banner advertising. By 1995 Yahoo! was adding a thousand new site listings a day and it took ten Silicon Graphics Indy Workstations to keep up with the 5 million hits per day. Now Yahoo! is an Internet portal site: the first site you see when you go on-line. It is getting into E-mail hosting, news, and shopping, and is becoming the home page of choice for web users.

Yang and Filo's editorial comment about the progress of the WWW at the Yahoo! site (9/98). "Every now and then we step back and marvel at how much the Web has changed since we started Yahoo! A lot of that change is reflected in the way Yahoo! has grown. What began for us as a fun way to collect our favorite sites has blossomed into a network of directories and features and news feeds, including everything from driving directions to sports scores, from live chats to the latest in online shopping. You can follow your stock portfolio online everyday. You can get weather updates for practically anywhere in the world, find old friends you haven't seen since grade school, download the latest games, get personalized news, place free classified ads, meet new friends...and the list goes on."

There are two ways to get listed on Yahoo! A site owner (webmaster) sends a request to Yahoo! to include the site in the Yahoo directory or a Yahoo! engineer finds the site from another source. In both cases Yahoo! has a person check out the site. If the web site passes muster, it is added to the directory. Yahoo knowledge engineers use what is submitted and their own experience and standards to index and categorize the site. It could take weeks to get listed on Yahoo.

A directory based search engine is preferred if you are looking for a mainstream item such as a company site. The directory sites excel when it is likely that the site you seek would have sent a request for inclusion to the directory in question. A search on a word like "maintenance" yields about 4000 hits from a directory site. The quality of the hits is high (relevant to your search needs).

The other type of search engine is a spider site such as AltaVista or WebCrawler. These sites continually search the web for new or changed sites. Spiders are also called robots or just bots for short. They use different strategies. Some create their index by reading the first few hundred words of the page. Some spiders also look for tags (called meta tags) that are imbed-

ded in the site. These tags are invisible to the viewer unless you set your browser to view the HTML code of the site. The Meta tags explain what key words the site should be indexed under. Some spiders catalog all of the links contained on a site and create an index of the links.

As a result of every page on the web being included in the search, spider sites return many more hits then directory sites. A search on a phrase like "maintenance" on AltaVista would yield almost 4,000,000 sites. These hits will be more complete but average a lower quality. This complete response is especially useful when searching a more obscure or very specific topic.

HOW TO USE A SEARCH SITE

Search sites' opening page (homepage) usually have a screen of banner advertisements and a search box. The search box is where the search words are entered. For simple searches, enter your topic or phrase (use quotation marks around phrases) and click the search button. Some engines have other options that restrict the search to a specific topic area or category such as business, computers, sports or entertainment, etc. If you were interested in Apple computers, a search of the whole World Wide Web for Apple would result in fruit, music (Apple Recording), geography (the big apple) and computers. Restricting the search to computers results in more relevant sites.

Search sites allow more complex expressions using Boolean operators (such as +, not, or, and). The rules are slightly different from site to site.

Usually there is an advanced area that describes the rules for complex searches. Most sites also have a help button for basic help and site usage. Before you use a search site for the first time find the help screens or read the FAQ file.

The key ingredient is knowing what word to use to find what you want. It takes some time to develop skill in this area. In using search engines, Tara Calishain says in her book *Netscape Guide to Internet Research* "The watchword for all of them, though, is flexibility. The more flexibility with which you can view a problem or answer a question, the better chance you'll have of finding the answer to your question or the information that you are looking for. The best tool in your Internet toolbox is still your brain."

SEARCHING THE INTERNET

Knowing something about the search engines will help you with your searches. The next section will compare the major search sites. The search itself is the key. Too broad of a search and you end up with thousands of sites or more. Too narrow a search and no sites will appear. Words and acronyms have very different meanings in different fields and different countries.

I looked up RCM (reliability centered maintenance) for a project. I thought RCM would produce a very targeted result. However, in some search engines, the number of useful sites was limited to 3 out of the first 25. Instead of the narrow result on reliability centered maintenance I found interesting sites on:

    The Royal Conservancy of Music
    Rubber City Manufacturing
    Arco (stock symbol)
    Royal Canadian Mounted (add Police)
    Many other results including those in other languages

As you can imagine, a search on plant maintenance yielded many green sites. In general, the more specific that the research topic is the easier it will be to find relevant sites. The more flexible the researcher you are, the more likely you are to find useful information. In other words how many different ways can you think of to describe the topic of interest such as preventive maintenance?

You could include maintenance, maintenance management, PM, preventive maintenance, predictive maintenance, tribology, lubrication, TPM, and RCM for starters. Searching on these key words would yield several million sites. Reviewing the top 25 in each area would give you many more ideas. Many search engines have a feature to allow such as "search for more of these."

Search sites are heavily-traveled portals to the Internet

Search Engine Name	Address or URL	# of weekly hits Jan 1998*
AltaVista	altavista.digital.com	6,764,000
Excite	excite.com	12,502,000
Hotbot	hotbot.com	2,703,000
Infoseek	infoseek.com	11,696,000
Lycos	lycos.com	6,787,000
WebCrawler	webcrawler.com	4,477,000
Yahoo	yahoo.com	26,726,000

*Relevant Knowledge Dec 29, 1997 to Jan 25, 1998

The search engines are big business. There are many mergers and acquisitions among in the search engines. Business and product advertisers see that the way to reach the demographically attractive Internet population is through their favorite search engine. Thus search engines are also engines of web commerce. As a result, many organizations are seeking strategic partnerships with one of the major search site owners.

The Walt Disney Corp. purchased a piece Infoseek for $75 million. The company sees the search engine as the portal to the whole Internet. Many believe (obviously Disney executives among them) whoever controls the portal's controls the Internet.

A company called Webpromote (http://www.webpromote.com) tracks the length of time it takes to get your submission onto each search engine. Webpromote also has a great service to help get your site ranked by the search engines. They send out a weekly update. This search engine status report is excerpted from the July 3, 1998 issue:

Yahoo!: The majority of sites achieving listing in Yahoo! will find indexing to be between four and eight weeks.

Infoseek: Sometimes instantaneous, other times indexing in days or more. Unpredictable. At times, not accepting submissions and other times, accepting but not actually indexing them. Since the introduction of "Extra Search Precision" (E.S.P.) on May 24, listings under popular one and two keyword search results now have a relevancy score of 100 percent. Through our research we have found that most pages with 100 percent relevancy have this score because they were personally reviewed prior to ESP kicking in.

AltaVista: They are still drastically limiting the number of daily submissions to only a few per day. Indexing remains between one and three days. With over 140 million pages, AltaVista maintains the largest number of web pages indexed. Wow! This massive number of pages offers searchers a greater chance of locating unusual or hard-to-find information.

On the next page the search sites are compared in two areas:

All of the sites are pretty easy to use for simple searches. Most also have extensive advanced capabilities.

	WC		Yahoo		Lycos*	Infoseek		Hotbot		Excite		AltaVista	
	GR	REL	GR	REL	REL	GR	REL	GR	REL	GR	REL	GR	REL
Maintenance Maintenance	35K	16	3K	23	8	1.5M	7	1.7M	12	.6M	12	4M	7
Manage	338	21	66	25	20	4K	22	12K	21	7K	25	18K	23
RCM	207	3	24	0	5	3K	19	15K	7	5K	6	39K	5
RCM2	7	2	0	0	4	19	5	273	4	10	4	200	6

GR- Gross responses called 'Hits'   *Lycos doesn't report the number of hits
REL- Number of relevant sites out of the first 25 hits   WC - Web Crawler

- ☐ Relevancy of the results: There is no advantage to turning up millions of sites if only a few are relevant to your search. Each search site has different strengths and weaknesses. You won't always know until you try a particular search on an engine. But by all means, try several engines for your more important searches. The sites are rated by the number of relevant sites out of the first 25 hits for each search.
- ☐ Gross number of hits: This measure tells you the total number of web pages that were found for that search. I don't think there is a meaningful difference between 1.7 and 4 million pages (or hits). It can make a difference, however, as your searches become more restricted or the topics more exotic.

SEARCH ENGINES:
AltaVista: http://www.Altavista.digital.com

Scope - Altavista is one of the largest and most comprehensive,engines indexing over 140 million Web pages. Scooter, the Web robot updates constantly. It is a full-text index that searches the entire HTML file and also gives the option for searching Usenet. AltaVista was designed to showcase the abilities of the DEC computer lines as high traffic servers. It does not index telnet, gopher or ftp. It is fast, handles millions of users, and has limited advertising.

Interface - AltaVista provides both a simple and advanced searches. In a simple search enter the word that you want AltaVista to look-up. Advanced searches allow limiting your search by date and to ranking results according to key words of your choice. You may also specify a format field to limit the search to hits that conform to the format (e.g. title, URL, host and links) or type of file (e.g., anchor, applet, image or text]. After the expression that limits the search put a colon followed by the search word.

Hit List for Maintenance Professionals	Number of Hits	Relevancy 1st 25 Hits
Maintenance	4,025,190	7
Maintenance Management	18,107	23
RCM (reliability-centered maintenance)	39,430	5
RCM2 (even more specific inquiry)	200	6

AltaVista hit more sites with maintenance by a factor of two. It was tied for second worst for relevancy of the site for our uses. This would be expected because of the complete site search technique used by Scooter. The maintenance management search showed very high relevancy or 23 out of 25 and also the largest number of hits.

Logic
- In simple searching, the default word is *or* between words.
- Advanced searching allows using full Boolean terms: *or, and, not,near.*
- Search for phrases searching by enclosing in quotation marks.
- Truncation is not automatic; in other words to get plurals or other additional variations at the end of words, add an asterisk (*) after a root of at least three letters to get up to five additional letters.
- + requires that a term or phrase be present; – denotes that a term not be present at all.
- Using upper case (capital) letters forces an exact match. Lower case will search both upper and lower case.

Results - The hit display shows title, URL, first two lines, date, size in bytes, and language. Duplicates are grouped under one title. Your output can be greatly improved by spending time learning the nuances of searching with AltaVista. Their on-line help pages have in-depth information.

Excite http://www.excite.com

Scope - Excite contains URLs and indexes two weeks of Usenet news and classified ads. It also offers a comprehensive, in-depth, browsable, hierarchical subject arrangement of 60,000 sources, NetReviews, a collection of brief Web site reviews. Excite's producers say that their NetSearch database is not padded by including the number of links embedded in indexed pages.

Interface - Excite provides simple and advanced interfaces. The advanced is basically identical to the simple, except it includes tips on how to improve your search by using Boolean logic when searching. Use both entirely separate databases for complete coverage: Net Directory for reviewed and Net Search for non-reviewed sources.

Hit List for Maintenance Professionals	Number of Hits	Relevancy 1st 25 Hits
Maintenance	631,187	12
Maintenance Management	7,790	25
RCM (reliability-centered maintenance)	5,153	6
RCM2 (even more specific inquiry)	10	4

Excite gave a reasonable number of maintenance sites with moderate relevancy (tie for 3rd). The maintenance management search gave the highest relevancy possible 25 out of 25 (tied with Yahoo!). Many sites on the list were useful and undetected by the other engines.

Logic

- + sign and – sign before words indicates must include or eliminate condition entirely.
- AND is the default, but it supports using OR and NOT, must be entered in capital letters.
- Searching by keyword or concept did not show a difference.
- Excite's search engine recognizes upper case letters on word beginnings as names.
- In a Boolean search, put names in quotation marks.
- Adding a ^ symbol and a numerical value to the end of a word increases its weight relative to the other search words and moves it higher in the results list.
- Terms are automatically searched as word prefixes or roots. Nesting term 1 with term 2 or term 3 is allowed.

Search features: Full text of WWW pages, excluding common stopwords; use of Boolean AND, OR, AND NOT; parentheses to group portions of Boolean queries together; "query by example" option finds pages similar to the search result you like best. "Concept-based search" (not available with above search features) attempts to retrieve documents with your exact query words and related words. Select "Advanced Query Language" for tips on how to fine-tune your search. The index is rebuilt approximately once a week.

Results - Excite is best used for finding widely discussed, mainstream topics. A chance to access similar sites is provided by "More Like this". The hit display includes title, URL, brief summary and "confidence" level with results shown in decreasing order of confidence. You may instead choose to display results grouped by Web site. Power search is available to further refine search results by entering specific variables.

Hotbot http://www.hotbot.com

Scope - Slurp, the Hotbot Web crawler has indexed millions of sites so far. The Inktomi search engine powers HotBot. It searches both WWW pages and Usenet news groups. It has no reviews or directories of sites, but has some unique search features as described below. It claims to be the first Internet search engine explicitly designed to permit easy scaling (if the traffic increases the engine can easily be scaled up to handle it) with the growth of the Web.

Interface - A choice is offered on whether HotBot searches your terms as individual words, phrases, a person's name or a URL. A pop-up menu lets you specify whether one or all of the keywords are found and in order or not. Using an expert result button allows further refinement of the search by date, media type, number of results per page, and location.

Hit List for Maintenance Professionals	Number of Hits	Relevancy 1st 25 Hits
Maintenance	1,739,537	12
Maintenance Management	12,315	21
RCM (reliability-centered maintenance)	15,597	7
RCM2 (even more specific inquiry)	273	4

Hotbot returned the second largest group of overall maintenance sites with an admirable 12 of 25 relevancy. The maintenance management search was more disappointing with lots of responses but relatively low relevancy (21 of 25 made it tied for 5th).

Logic

- ☐ Two optional buttons for simplify and modify are available. Modify narrows the search for more accurate results. It allows specifying phrase searching or you can use quotation marks around a phrase.
- ☐ A "must not" button eliminates a term.
- ☐ A "should" selection tells the search engine to place more emphasis on that term.
- ☐ At this time Boolean searches are supported only through the popup options.

- Case is insensitive in HotBot except the very rare instance of mixed case within the same name.
- Date searching gives an option of all in the database, options for documents older or newer than a specified date, or content posted within the last few days or months.
- Searching for specific technologies on pages that contain Java, JavaScript, VRML (3 D), Acrobat, or smiley is possible. Also, you can search for file type by suffix such as GIF.
- You may search by .edu to get educational institutions or .uk to get United Kingdom or by actual domain name of a specific Web site; (www.hotbot.com).
- The Geoplace button allows searching North America, South America, Asia, or Europe for Web servers in those locations. The default is anyplace.

Results - Relevancy of results is based on several factors: word frequency, number of times words appear in the title, document keywords, etc. Duplicate sites are listed with only one title, but all URLs are given which allows accessing the document on a server closer to you. You may request search results in full description, brief description, or URLs in increments of 10, 25, 50, 75 or 100. Searching for a person returns results based on several different combinations of a name. Searching a URL returns documents that contain that URL in either a link or document text..

Infoseek http://www.infoseek.com

Scope - With millions of sites in full-text, this is one of the best tools for comprehensive searches. For speed, relevancy and currency, it is hard to beat. The Infoseek Guide database counts unique Web pages, not the URLs (or links) mentioned in those pages. Infoseek covers WWW pages, Usenet news groups, Infoseek Select Sites (reviewed), directories of companies, E-mail addresses, news stories, and FAQ's.

Interface - A simple search interface with a pull-down menu (which gives you access to the advanced capabilities)lets you limit your search to the World Wide Web, gopher, FTP, Infoseek Select sites, Usenet newsgroups, company directory, e-mail addresses, timely news or Web FAQ's (frequently asked questions.)

Hit List for Maintenance Professionals	Number of Hits	Relevancy 1st 25 Hits
Maintenance	1,580,754	7
Maintenance Management	4377	22
RCM (reliability -centered maintenance)	3073	19
RCM2 (even more specific inquiry)	19	5

The maintenance search turned up the third leading number of sites but mediocre relevancy (tied for last). Once you have selected a main topic, you can narrow the search before you use the search index by selecting a subtopic.

Logic

- Boolean *or* is the default, but Infoseek allows variations of *and, not, adjacent,* and *near.*
- Search for phrases by enclosing in quotation marks.
- A hyphen between words locates those within one word of each other.
- Brackets around words will retrieve words within 100 words in any order.
- Use a + sign to ensure inclusion of a word or phrase and a - sign to exclude a word or phrase.
- Searches are case sensitive; capitalize proper names and insert a comma between different names.

Results - Up to 100 items are displayed per search in the result list, ranked by relevancy. A display includes the document title, URL, size in bytes, relevancy score, and first three lines of text and computer-generated summary. Some results include a list of related topics and news groups. Fewer duplicates are retrieved from Infoseek than any other search engine. When you locate a site that seems relevant, you may improve your results by selecting the Similar Pages command. Also available is Infoseek Select: 12 subject headings giving very brief information on reviewed World Wide Web pages. Infoseek endeavors to be selective in the pages it indexes so as to reduce the number of duplicate and irrelevant pages retrieved. For best results, use the search tips.

Lycos http://www.lycos.com

Scope - One of the largest search tools on the Internet, Lycos covers over 90% of web sites. Millions are fully indexed: title, URL, headings and sub-headings, first 20 lines of text, 100 most weighty words, etc. Thus Lycos, primarily searches URLs and abstracts. However, it does search gopher and ftp sites as well as five million binary files: GIF, JPEG, and MPEG for images, sounds and movies. Also offered are WWW reviews by Point Communications and search of a subject-based directory of the most popular pages from the Lycos database called a2z. The directory can be searched by keyword and a brief description is provided for each site. Lycos merges the results of its continuous WWW sampling into its catalog on a weekly basis.

Interface - Lycos offers both a simple and an advanced search. The advanced interface includes options to modify the search logic, the amount of detail in the results display, and the number of results displayed per page. It ignores common stopwords; it searches the title, headings, links, keywords, and first 20 lines of Web pages.

Hit List for Maintenance Professionals	Number of Hits	Relevancy 1st 25 Hits
Maintenance	Number of sites not listed	8
Maintenance Management	Number of sites not listed	20 (many duplicates)
RCM (reliability-centered maintenance)	Number of sites not listed	5
RCM2 (even more specific inquiry)	Number of sites not listed	4

Logic

☐ Boolean *or* is the default.
☐ The advanced interface allows the searcher to specify *and, or, not* between words.
☐ Advanced search also allows a searcher to specify whether a match should be loose, fair, good, close, or strong.
☐ Truncation is automatic; you may use a period after a term to turn off stemming or use $ to manually truncate terms.
☐ Using a minus sign - before a term eliminates it from the search.
☐ Using a plus sign + ensures a term will be included.
☐ Also allows narrowing the search to sounds or images.

Results - Three output options: just the links; standard description which includes title, first two lines of text, URL and confidence level score; detailed description which gives ranking, outline, abstracts, URL, size and date, and the number of links in document to outside sources. Provides extensive search tips.

WebCrawler http://webcrawler.com/

Scope: WebCrawler allows document title and content searches of its sub-mission and robot-constructed, content-based database. The database consists of both explored and unexplored Web pages. Unexplored web pages are derived from links on explored pages. Brian Pinkerton wrote Web Crawler while a student at the University of Washington. The first release was April 20, 1994.

Hit List for Maintenance Professionals	Number of Hits	Relevancy 1st 25 Hits
Maintenance	35,198	16
Maintenance Management	338	21
RCM (initials for reliability centered maintenance)	207	3
RCM2 (even more specific inquiry)	7	2

Logic

- □ Words automatically stripped of their endings
- □ Allows natural language queries
- □ Use of Boolean AND, OR, NOT
- □ + before word means that the web site must have the word
- □ - before word means that the web site must not have the word
- □ Phrase searching; adjacency; parentheses for grouping searches
- □ Proximity searches.

Results: Search results are returned in order of decreasing relevance in an unannotated, easy-to-browse list. Options including detailed and short formats. Useful search tips and examples are available.

Yahoo! http://search.yahoo.com/bin/search/options

Scope: Yahoo! searches by WWW pages, E-mail addresses, and Usenet newsgroups. After you have specified keyword(s) inside the query box and clicked on the search button, Yahoo will search through the five areas of its database for keyword matches. The five areas are:

- □ Yahoo! Categories.
- □ Web Pages
- □ Yahoo! Web Sites
- □ Yahoo!'s Net Events
- □ Most Recent News Articles.

500,000 pages indexed into 25,000 catagories. The file can be searched (using key words) or browsed by using the category button. You can search a directory such as Yahoo! But, you can also browse through its rich hierarchy of information: a vast collection of categories and sub-categories created by people, not computer programs. And by browsing, you can have in front of you a good, perhaps complete listing of all the sites that cover a particular subject-just look for the category.

One important difference between search engines and directories is that directories have structure. You can navigate this structure and peruse the information contained within it, making choices as you go along. With a directory such as Yahoo!, you can choose to navigate your way to a particular category--and once there, know that you haven't missed anything (excerpted from the Yahoo! tutorial).

Hit List for Maintenance Professionals	Number of Hits	Relevancy 1st 25 Hits
Maintenance	3416	23
Maintenance Management	66	25
RCM (reliability-centered maintenance)	24	0
RCM2 (even more specific inquiry)	Switched to AltaVista	

Logic

　　　□ Matches can contain all or at least one of the search terms, selected substrings, or whole words.
　　　□ Divides the hits into categories (setup by Yahoo!) for subsequent searches.
　　　□ Select the number of matches displayed per page.
　　　□ Yahoo! is not case sensitive.

Example of Yahoo! advanced search screen options:

Search: Yahoo! Usenet E-mail addresses. For Yahoo! search, please use the options below.Select a search method:
　　　　　　Intelligent default
　　　　　　An exact phrase match
　　　　　　Matches on all words (AND)
　　　　　　Matches on any word (OR)
　　　　　　A person's name
　　　　　　Select a search area:
　　　　　　Yahoo Categories
　　　　　　Web Sites
Find only new listings added during the past ( ) years
After the first result page, display ( ) matches per page

Results: Yahoo! returns a list of categories that match your keywords, end-sites that match your keywords, and names of categories where those end-sites are listed.

Comments: Yahoo! searchers may use the name of the categories included in search results to go seek other similar items.

SMALLER SEARCH ENGINES

Galaxy Search http://galaxy.tradewave.com/search.html

Databases: WWW pages, Gopher titles, and Telnet resources from the Hytelnet hypertext telnet database.

Search features: Full, title, and text searches, retrieval of ANY or ALL search terms, automatic truncation.

Results: Output length (annotation of results) may be selected.

Comments: Use the name of the subdirectory included with each retrieved item to go seek other similar items in Galaxy. Provides some search tips.

DejaNews http://www.dejanews.com/

Databases: Archived Usenet news groups "ranging back to March, 1995."

Search features: Select "Power Search" to take advantage of the full range of search options including the Boolean operators AND, OR, AND NOT, and NEAR. Other features are phrase, proximity and field searching.

Results:
Comments: Updated daily. An extensive help index is available.

The Open Text Index http://index.opentext.net/

The Open Text Index indexes the full text (even common stopwords) of Web pages.

Databases: WWW pages, Usenet news groups (Deja News)

Search features: Choice between Simple Search and Power Search; Boolean AND, OR, BUT NOT, NEAR, FOLLOWED BY; limit of search to certain parts of Web pages such as URLs, titles, headings.

Results: Special features allow users to view their search terms in context and find pages similar to the ones they like best.

Comments: Open Text does not truncate pluralized\words. Be sure to use the search tips.

Savvy Search http://www.cs.colostate.edu/~dreiling/smartform.html
SavvySearch is a multi-threaded tool, designed to take your query to several search tools at once, gather the information, and return the results.

Databases: WWW pages

Search features: Using a common interface for several search tools prevents the use of certain special search features available at some sites.

Results: Option to select length of format. Results are arranged by name of search tool. Presence of annotation varies depending on the original search tool.

Comments: Due to its popularity and the problems inherent in contacting multiple search tools, SavvySearch can be difficult to reach during the day.

Thanks to the Kansas City Public Library, and the Northwestern University Library and the help screens from the search engine sites for the reviews and compilation of information.

Meta Search engines

The Meta search engine submits your search to several engines at the same time and returns the results from all the engines. The Meta search engine usually filters out duplicate URLs. Using one of the Meta search engines mentioned below might speed a search or insure that all appropriate sites are included.

www.dogpile.com Dogpile offers efficient search of 23-search engines

http://m5.interence.com/ifind/ Searches several engines and sorts and eliminates duplicates:

http://www.mamma.com Mother of all search engines.

ONLINE REFERENCE LIBRARIANS

In a Reuters story by Michelle Rafter the digital replacement of the research librarian is introduced. In the days before the Internet, if you wanted to learn more about a subject a good place to start was the librarians at the desk. If they did not know the answer to your question, they always knew where to look. Now, with the Web evolving into the world's largest repository of information, on line companies such as Ask Jeeves (http://www.askjeeves.com), InfoPlease (http://www.infoplease.com), Answers.com(http://www.answers.com) and Electric Library(http://elibrary.com) are attempting to duplicate online services of flesh-and-blood reference librarians. Expect to see more of these reference services soon.

Possibly the best feature of these new reference desks is their ability to understand queries written in plain English. That means searchers can ask simple questions rather than using arcane Boolean search terms that are standard operating procedure in Web directories such as Yahoo and Excite. Like the search engine sites most of these reference services are free. Companies support their operations through advertising and partnerships.

The best was Ask Jeeves, which returned detailed answers for two of the three test questions. Named after British novelist P.D. Wodehouse's all-knowing butler, Ask Jeeves was created 15 months ago by a private Berkeley, California, company of the same name and recently spun off a separate children's' service called Ask Jeeves for Kids (http://www.ajkids.com).

On the first question (Who was the 16th president of the United States?), Ask Jeeves offered an almanac entry on Abraham Lincoln that included a short biography, familiar quotations, inaugural addresses, a photo and links to famous speeches and Mary Todd Lincoln. It also delivered pointers to related information such as U.S. presidents and history. In Ask Jeeves for Kids children create some content.

SUPERSITES

Supersites are the diamonds of the Internet. When you find one, carefully bookmark it so that you can always find it again. For some topics the supersite blasts open the WWW and gives you access to hundreds of resources. Supersites save you hours of, sometimes, fruitless searching. Someone took the time to do research and assemble on his or her site links to related sites.

Supersites are web sites that have assembled a group of sites related to a single topic, user profile, interest group, or location. These sites are very important for new users in a field because much of the time on the Internet is wasted building the same list that the webmaster of the supersite already built.

There are many types of supersites. Some charge fees to establish links. Magazines are examples of fee-based supersites (all of the advertisers are represented by links). Industry associations frequently sponsor supersites of member organizations. Some sites are works of dedicated individuals that are committed to a field. The government also funds research where the deliverable is a supersite with links to resources on a particular topic.

The first supersite below is a favorite. It was the first maintenance-oriented I found during my first few days of surfing.

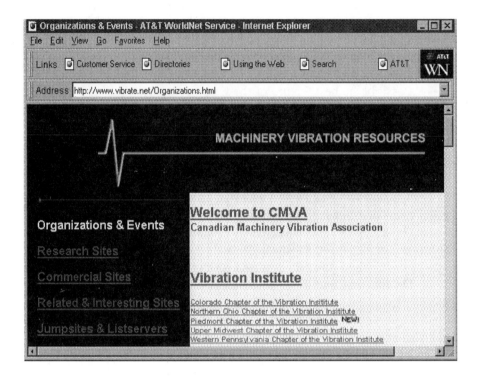

The next three supersites are special types of sites. Both Thomas Register and Industry.net are subscription supersites. Companies pay to be listed in these sites. In some cases the fee is comparable to the cost of running a full-page ad in a major publication.

Most magazines qualify as supersites because they provide links to all of their advertiser sites. Magazines are such a rich treasure trove for maintenance management that they have their own chapter in this book. If I wanted a quick introduction to a new field, I would look up the trade magazine's sites and dig right in.

The venerable Thomas Register has its own site on the World Wide Web. No introduction to the Internet would be complete without visiting http://www.thomasregister.com.

A search of Thomas Register for press brakes came up with three vendors. One vendor had an on-line catalog. The information card for that manufacturer follows on the next page.

Another service can be found at http://www.industrynet.com. All kinds of companies have joined together on this service bureau. The site provides many links to company home pages. Industry.net's Data Depot is a gateway to comprehensive technical industrial products and components information. Make design and specification decisions using complete technical product descriptions with Industry.net's exclusive Industrial Catalogs at http://www.industry.net/pm-mktg/icdescription.htm. When you join you can

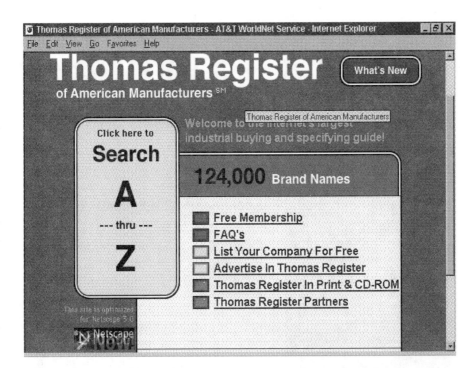

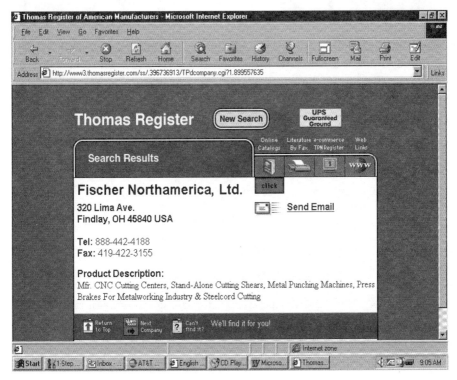

access catalogs from over 15,000 manufacturers and distributors of industrial products and services. The Industrial Catalogs have been subject indexed and cross-referenced enabling you to quickly search and FIND the company or product you are interested in.

In a recent communication to members they layout the future offerings. Future additions to Industry.net's Data Depot will include online ordering of more than 200,000 technical standards, specifications, and reference documents; downloadable industry standards; and key supplier information on more than 12 million parts.

Industry.net's Reference Shelf provides fingertip access to many reference tools and resources that engineers use on a daily basis.

☐ Make sure your technical documentation is in compliance with current industry standards with the online Drawing Requirements Manual.

☐ Enjoy convenient, fingertip access to frequently used (but often forgotten) engineering equations and formulas.

☐ Apply NASA technology to your commercial applications and product development.

☐ Stay abreast of changing federal and state safety and compliance regulations.

☐ Catch up with previously published Industry.net information in our Archives section.

□ Keep pace with manufacturing technology with the Integrated Manufacturing Technologies Roadmap (IMTR).

Manufacturing Online (http://www.chesapk.com/~chesapeake/ offers Internet users access to demonstration software available for upload, links to manufacturers and extensive useful lists.

http://www.techexpo.com If you want to visit many manufacturers at one time face to face, you might consider visiting a trade show. A site called TechExpo tracks trade shows. TechExpo also has files on thousands of conferences on technical, engineering and science topics.

http://www.Part.net/ is a linked resource for organizations wanting to locate vendors of electronics parts. Parts vendors for 12 million parts used by the US Department of Defense. PartNET uses typical hyperlinks and Internet-type search engines to search for a particular commodity. PartNET is available for vendors in several ways including storefront (where anyone can buy and everyone sees the same catalog), supply chain (where many of your unique vendors are in a virtual mall), or dedicated buyer sites (where large buyers have customized sites with the parts they buy and the prices they pay).

http://www.industrialcourcebook.com The Industrial Sourcebook has been a mainstay of research for parts and vendors in Canada for many years.

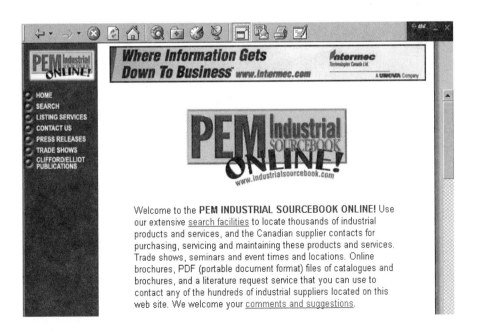

SUPERSITES WITH LINKS TO OTHER SITES

http://mtiac.iitri.com/ Resource with links to manufacturing resources throughout the net. It's aim, according to *Manufacturing and the Internet* "is to become a comprehensive listing of Internet resources in defense manufacturing as well as a reference for related resources at universities, in industry, and at other government organizations."

http://mfginfo.com/home.htm The manufacturing information net. It has a search engine available to find manufacturing resources throughout the Internet, software to download, classified ads, and on-line discussion areas.

http://www.industrialview.com The brainchild of longtime maintenance professional Dick Wilson and his son; gives Northwestern United States MRO vendors a forum for showing their wares. A search facility helps you find useful information.

http://conduit2.sils.umich.edu/ The Manufacturing Information Solutions Web site designed for small manufacturers.

http://www.steel.org/ The home of the American Iron and Steel Institute. They sponsor research and have special interest groups in automotive, appliances, and construction topics.

http://www.mmf.com/metal/ The Metal Machining and Fabrication International Internet Directory has links to vendors, consultants, finishers, equipment vendors, trade shows, tips, equipment for sale, materials for sale and a host of other capabilities. This is a particularly complete site for visitors seeking any kind of metalworking. It includes good links to colleges that have departments of engineering specializing in metalworking.

http://www.asisonline.org The site for the American Society for Industrial Security. Click the network button to find local chapters of the association and related sites of interest.

http://www.miningusa.com Mining Internet service is an Internet storefront to the mining industry. Its Table of contents includes:

Mining Associations	News Links	Company Store
Companies	Mining History	Conference Room
Consultants & Services	Research	Government Agencies
Mining Suppliers	Tourism	Other Links
Mining Publications	Educational	
Employment	Environmental Links	
Mineral Law	Calendar of Events	

http://www.sweets.com/ One of the oldest catalog of catalogs is Sweets. Architects have used Sweets to find vendors and specifications for all parts of the building trade.

Division 01 - GENERAL DATA	Division 09 - FINISHES
Division 02 - SITEWORK	Division 10 - SPECIALTIES
Division 03 - CONCRETE	Division 11 - EQUIPMENT
Division 04 - MASONRY	Division 12 - FURNISHINGS
Division 05 - METALS	Division 13 - SPECIAL CONSTRUCTION
Division 06 - WOOD & PLASTICS	Division 14 - CONVEYING SYSTEMS
Division 07 - THERMAL & MOISTURE PROTECTION	Division 15 - MECHANICAL
Division 08 - DOORS & WINDOWS	Division 16 - ELECTRICAL

Designed exactly like the catalog, it allows searches by division (similar to the way that a specification for a building would be written). The Internet version also allows search by keywords, company, and trade names. This site is very useful for major building rehabilitation and construction.

http://www.industrylink.com/ A supersite with hundreds of links to resources in specific industries from chemical to mining to automation.

http://www.efn.org/~franka/MMeasier/TIPS.html This site is a yellow pages of maintenance management resources including a very complete listing (with links) to CMMS organizations, maintenance magazines, tips, jokes, and other important contributions to your day.

http://www.sunbeltengineering.com/mis_lnk.htm This is the site of a consultant with useful links to many unique resources. The site includes government, manufacturer, software and organizations links.

5

MAINTENANCE MAGAZINES, PUBLISHERS AND ASSOCIATIONS

Among the greatest resources of the Internet are the web sites by the major maintenance magazine publishers. These web-based magazines (called webzines when they are solely on the web) have several advantages over their print counterparts. The home page for Maintenance Technology is **http://www.mt-online.com**.

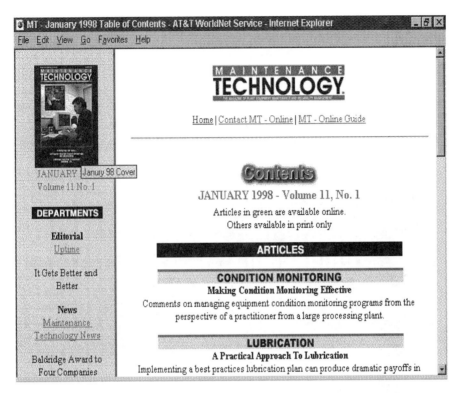

The web version doesn't take up space in your office and it doesn't consume trees! The publisher can leave an archive available for research. Links are available to all of the advertisers. Once you hyperlink to the advertiser, a world opens up. The advertisers can have their entire catalogs and price sheets supplemented by videos, photographs or drawings. Quotes and proposals can be just a click away.

Some of the more sophisticated magazines send you E-mail with the contents of the new issue. When a webzine or product information is sent to your E-mail account the technology is called "push." Push is one of the most powerful abilities of the web. New HTML-enabled mail programs (such as Outlook '98) will allow color magazines to be sent as E-mail (the magazines will still stack up in E-mail like they currently do on the floor- but at least it will be harder to trip over an E-mail stack!).

The table of contents for the January issue of Maintenance Technology follows. Following Maintenance Technology is the Plant Services' home pages. Each is a powerhouse in the maintenance field.

Internet magazines have several major functions. In some cases like their print counterpart they serve the same functions as paper magazines, the magazines should:

1. Keep the reader current with new ideas and trends and introduce them to the people of their industry.

2. Provide access to product information, particularly for new products, by printing news releases from vendors and running informational advertising.

3. Have excellent articles that, when read over a period of time, build expertise in the field.

4. Have a section of job openings in the field.

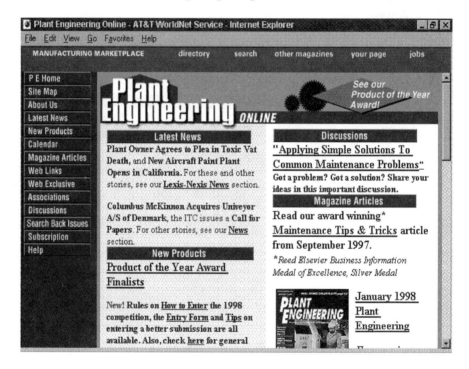

Plant Engineering Magazine http://www.manufacturing.net/magazine/planteng has the highest circulation of the popular pure maintenance magazines. As you can see from the previous page the web site is full of resources. At publication time some of the links are not complete and some of the capabilities were still skeletal.

CAPABILITIES TO LOOK FOR IN WEB BASED MAGAZINES

What new and different capabilities should be available in web based publications? In addition to the above, we should look for to following:

1. A searchable archive of past articles indexed by topic or author to allow research. If you were going to install a new CMMS, it would be very useful to be able to read a year or two's articles on the topic.

2. If the web magazines are sent to you by E-mail, then the company should send short messages to alert you to an impending problem or short-term opportunity. I receive a daily morning newspaper. The company also sends me alerts about any major breaking stories.

3. The newspaper I receive each morning by E-mail has another aspect impossible on paper (unless you can pay thousands): customization. I selected the categories and topics of stories when I subscribed for my personal version of the newspaper. My morning newspaper is personalized to cover only the issues of interest to me! Because there is no printing cost, the newspaper is totally supported by ad revenue. I don't have to pay for it. Again, any publication that uses this technology is called 'push'. You will be hearing a lot more about this in the next few years.

4. A powerful capability of the WWW (not yet used to my knowledge by the maintenance publishers) is the ability to transmit videos and sound. The barrier today is the time taken to download the files. Imagine in the near future an article on vibration analysis with some well-chosen videos and a few vibration analysis advertisements with a live video showing the

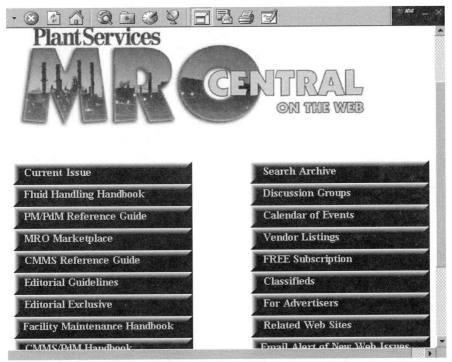

Plant Services has an excellent web site with a good archive of past articles.

product in action!

5. Web-based magazines have links to advertising sites. You can satisfy your curiosity about a product immediately, in detail, by surfing over to the manufacturer or vendor's site.

6. Magazines usually have useful links to related sites for additional information, research, and information. As mentioned in chapter four, most magazines are also themselves supersites.

7. Web-based magazines don't take up space, or use up trees.

SAMPLING OF MAGAZINES OF INTEREST TO THE MAINTENANCE PROFESSIONAL

http://www.csemag.com. For the engineers out there, there are many sites to visit. Start with Consulting-Specifying Engineer magazine.

http://www.amtonline.com/ The magazine for airplane maintenance. In many ways, airplanes are the most sophisticated end of the maintenance business. RCM and RCM2 both have their origins in the aircraft field.

http://www.fleetowner.com/default.html The web version of one of the more popular fleet maintenance magazines. Many useful resources.

http://www.impomag.com/ Similar to the print edition in that it is easy to use, articles are interesting and timely. All the issues feature archives, hot topics, a stockroom, a calendar, a forum, links to manufacturers, and many other useful capabilities.

http://www.truckfleetmgtmag.com/ This is the web site of a network of services for the truck maintenance professional.

http://evolution.skf.com/gb/eng-main.asp The site for the internal magazine for SKF. If you work on the mechanical side of maintenance, this site includes interesting information.

http://www.tfmgr.com Today's Facility Manager covers buildings and facilities

http://www.newequipment.com. A generalized resource available from the New Equipment Digest magazine. The site has more than 5000 descriptions of industrial products in various categories. The site can fax your request for additional information directly to the appropriate manufacturer. You can also sign-up for an E-mail service which sends you notices when new products come out.

http://www.sensorsmag.com/ SensorsWorld Link is sponsored by Sensors magazine. It will link you to vendors, articles, and a FAQ file about sensors. Registration is required.

http://popularmechanics.com/ What self-respecting maintenance professional hasn't spent the odd hour or more looking at Popular Mechanics magazine? Now we can visit the site on the web!

Are you interested in logistics and warehousing? On this page magazines were displayed on **http://www.cap-ai.com/library2/magazin.html**

Logistics and Warehousing-related Magazines

Title of Magazine	Available from	Cycle of publication
Automatic I.D. News	Automatic I.D. News (USA)	Monthly
Distribution Magazine	International Fleet Management Consultants, Inc.	*
Industrial Engineering	The Institute of Industrial Engineers (USA)	*
Industrial Equipment News	Thomas Publishing Co. (USA)	Monthly
Industrial Handling & Storage	Trinity Publishing Ltd. (UK)	*
Industrial Maintenance & Plant Operation	IMPO (USA)	Monthly
Journal of Business Logistics	Council of Logistics Management, (USA)	Biannually (Spring / Fall)
Logistics Billboard	PNL Publishing	*
Logistics Business Magazine	Interactive Business Communications Ltd. (UK)	Quarterly
Logistics IT Magazine	Interactive Business Communications Ltd. (UK)	*
Logistics Management	International Fleet Management Consultants, Inc.	*
Material Handling Business	Material Handling Engineering (USA)	Quarterly
Material Handling Engineering	Material Handling Engineering Magazine (USA)	Monthly
Materiall Handling Product News	Material Handling Product News (USA)	*
New Equipment Digest Magazine	New Equipment Digest Magazine (USA)	Monthly
Plant Engineering	Plant Engineering Magazine (USA)	*
Transportation & Distribution	Transportation & Distribution Magazine (USA)	*
Warehousing Management	Warehousing Management Magazine (USA)	*

MAINTENANCE MANAGEMENT PUBLISHERS ON-LINE

In addition to magazine publishers, book publishers of interest have excellent sites. Industrial Press has been an active resource for the maintenance department with Machinery's Handbook ever since the first pamphlet came out in 1883. They have an extensive backlist of maintenance books for management,

engineering and practice. Industrial Press also has the distinction of being a super supersite. Most of the links listed in this book are accessible through the IP site. (**http://industrialpress.com**).

The World Wide Web allowed the development of special purpose companies that are both problem solvers and vendors to very specific markets. Because of the global reach of the web and the modest cost of starting up small boutique shops are viable. One of these boutique publishers is devoted to the advancement of the maintenance field. TWI offers a variety of resources for the maintenance professionals including books, links, and CD-ROM products (**http://www.twipress.com**).

A FEW OTHER PUBLISHERS OF INTEREST TO MAINTENANCE PROFESSIONALS

http://www.conference-communication.co.uk/books/ A British site with links to useful books on maintenance management.

http://www.mcgraw-hill.com/books.html One of the largest publishers and one that has a significant maintenance presence is McGraw-Hill. You can access their bookstore and also their individual publishing imprints.

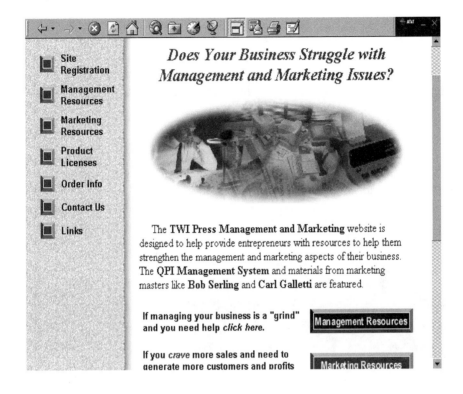

http://www.prenhall.com/ Prentice-Hall is one of the leading publishers in the maintenance management field.

http://www.mcp.com/ Macmillan press is a leading publisher in the technical and computer field.

BOOKSTORE PAR EXCELLENCE

http://www.amazon.com You can't spend any time on the web without running into a banner for a unique bookstore called Amazon.com. Some people might shy away thinking they are not interested in rain forests or South American warrior women. Amazon.com is a virtual bookstore and a 'must see' site. It has identified millions of books on the shelves of publishers, distributors and major retail vendors. In most cases it provides these books in a few days by UPS.

Amazon.com's approach to advertising is unique. For example, when you search on the civil war using one of the search engines, the Amazon.com's banner appears on the results page. It reads 'for books on this topic press this button'. If you click the button, a list of civil war books comes up on the screen. A few clicks later (and a credit card) and books are on their way to you. Almost any topic that you search leads to an Amazon.com a book list. Amazon.com is available directly or through www.maintrainer.com.

Since Amazon.com became so popular the two major bookstore chains (Barnes & Noble and Borders) have followed suit.

TRADE ASSOCIATIONS

One of the leading associations for pure maintenance is the Association for Facilities Engineering (originally called the American Institute of Plant Engineering). At a recent meeting there were schools, automobile factories, water utilities, hospitals, chemical plants and hundreds of others represented. This association crosses industry and maintenance type lines. Members include every industry that has a substantial maintenance or plant engineering activity. http://www.afe.org

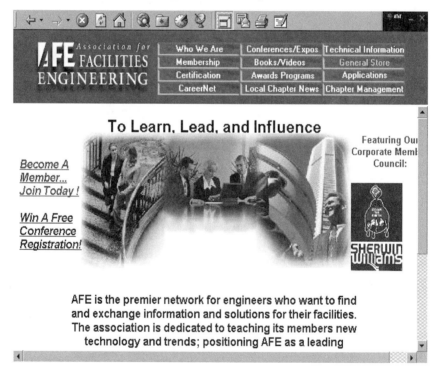

Other associations to visit are:

http://www.baqa.org The home page for the Building Air Quality Alliance BAQA.

http://www.asme.org/index.html The American Society of Mechanical Engineer's web site has significant capabilities for members. The most interesting feature is access to a complete library of free downloads of software, journals and other publications.

http://building.com/index.html Building Industry Exchange Foundation

http://www.camc.ca/camcenglish/index.htm The Canadian Aviation Maintenance Council has interesting information for anyone on how maintenance is conducted in the many aviation business. The related site **http://www.groundeffects.org** has research into avoidance of mistakes.

http://informs.org/Conf/WA96/TALKS/WB28.html Institute for Operations Research has a section devoted to maintenance. Some of the postings look pretty interesting:

> **WB28.1** Designing a Predictive Preventive Maintenance Program
> Popova, Wilson
> **WB28.2** Maintenance of a Markov Modulated Reliability Model S.
> Ozekici, M. Sevilir
> **WB28.3** Probabilistic Design of Rotors: Minimizing Imbalance
> Shortle, Mendel
> **WB28.4** A Model for Autonomous Maintenance Investments
> McKone, Weiss

http://www.boma.org Building Owners and Managers Association BOMA

http://www.ewi.org/ewi/ Edison Welding Institute is a membership institute dedicated to welding and joining technology on the factory floor.

http://www.iienet.org/ Institute of Industrial Engineers (their publishing arm publishes *Manufacturing and the Internet*) has excellent resources for maintenance professionals who are interested in re-engineering the process and techniques of the maintenance effort.

http://www.hsb.com/pcm/mimosa/mimosa.html Machinery Information Management Open Systems Alliance (MIMOSA)

http://www.smrp.org Society for Maintenance and Reliability Professionals

EQUIPMENT, COMPONENT MANUFACTURERS AND INDUSTRIAL DISTRIBUTORS

Much of the power of the Internet for the maintenance world depends on how well the OEMs develop their web capability. Right now web site excellence is very spotty. National Semiconductor is an example of a site

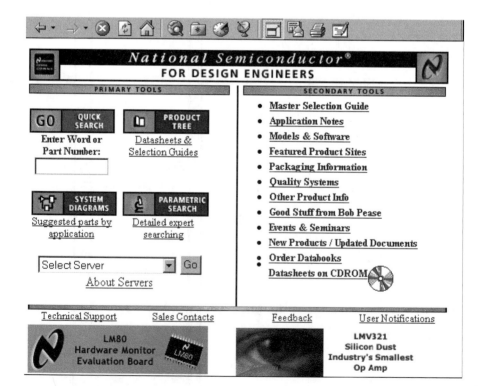

with good information for users. Most sites are taking baby steps by having just put catalogs on the WWW, (which is of some use). There is a wide world beyond catalogs. The potential of the Internet is to lower costs while increasing the level of service and staying closer to the customer.

The SKF homepage (http://www.skf.com) is typical for the brochureware or catalog site. There maybe useful maintenance information but it is buried behind pictures and news stories. As maintenance professionals, we need hooks into these megasites that deliver us to some useful place. *Evolution* (the SKF magazine) is now available on the web.

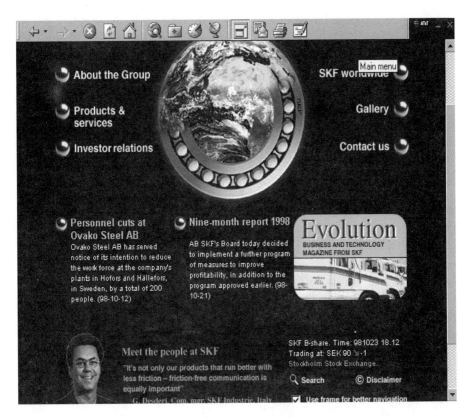

A 1997 survey by Advanced Management Research showed that 71% of manufacturers already use the Internet to communicate with both their sales forces and their customers. An additional 11% are expected to jump on the bandwagon in 1998. Many reported that developing the Internet communications channel is a major priority for the next year. It is not premature to say that if an OEM is not on the Web by the beginning of the year 2000 it is going to be locked out of most of sources of business.

Thomas Register and VISA conducted a survey of industrial purchasing on the Internet and found that, out of 2000 companies surveyed, about 10% used the Internet for purchasing by the beginning of 1998. By the end of the year, 21% of the purchasing executives expected their organization to be using the Internet (reported in MT 4/98 P7).

HOW DO YOU FIND THE MANUFACTURERS

There are three major ways to find manufacturers. The most common way is to use the search engines. One advantage of the search engines is that the search can be set up to return either very specific (a single manufacturer) or very general results (a category of manufacturers that might compete). For example, you can search on the category of the product, such as bearings, or you can search on the specific vendor, such as SKF. A wider category search will also retrieve competitive bearing sites for comparison-shopping.

The second common way to locate a manufacturer is to visit one of the industrial supersites like Thomas Register, Industry Net, Sweets or an appropriate trade magazine site. Most trade journal advertisers maintain web sites and most magazines have links to their advertisers' sites.

The third way is low tech. Save advertisements and articles that have useful web sites and root through them whenever you want a particular manufacturer. Or look them up the next time on-line and bookmark the useful ones. You can also try the company name or initials in the URL window of your browser (that sometimes works!) and hit enter. For example, try gm.com for General Motors or ibm.com for IBM.

Maintenance management professionals have serious business on the Internet. In the near future, if not already, information from Web sites will minimize downtime, reduce breakdowns and, maybe even save lives (all by having critical information in the hands of someone who can use it when it will make a difference).

We expect the equipment manufacturers to invest enough money that the capabilities listed below are increasingly available. The drive for these improvements will come from equipment users, such as you, making requests and getting promises from the vendors. In other words, our repeated requests turn into next year's capabilities. The kinds of uses that we are seeking from OEMs include a good deal more then catalogs. Some of the other capabilities include:

1. Access to catalog cuts, specification sheets, and dimension drawings. This is a very useful capability when you have a specific question for a report, proposal or project. Dimension drawings and complete descrip-

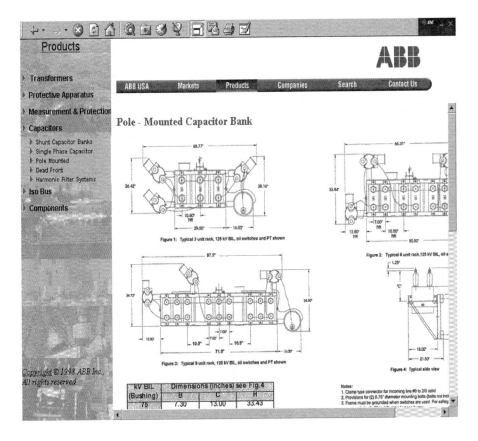

tions/specifications allow you to answer important questions like, will it fit? Catalogs are available all day and night so you can work at your convenience. This is a common capability.

Don't get me wrong catalog sites have their uses. Currently the most common use of maintenance for the Internet is to get basic information for a project quickly. OEM catalog sites are perfect locations to gather information. A catalog site has modest goals: it just wants to duplicate the catalog. Usually, as in the example above, the site adds communications capability. The ABB catalog site has drawings and tables to use when sizing one of its products. (http://www.abb.com/abbus/Corporate/Product/t&d/index.asp)

2. The ability to get quotations quickly and communicate with sales departments. Sometimes just getting through endless voice mail to someone who can quote you is time consuming. In one case, I needed a quote on a high-speed scanner. I e-mailed my request to the company, within hours the quote was in my mailbox. I saved four or five phone calls and an hour on the phone. Expedited quotes are a common capability.

3. OEM web sites can provide access to engineering drawings, particularly for obsolete equipment that is no longer supported. This capability is underutilized but it is an excellent use of the Internet. It is a low priority for most OEMs although it would be an extremely low cost way to continue to support customers of older units.

4. Access to current O & M manuals and maintenance tips. Baby steps have been taken in this area. The best manufacturers are the software houses. I'm stretching the point calling them OEMs, but they have tips sections and access to some manuals for reading or download. There is a universal portable document format (.pdf) from Adobe called Acrobat. Documents and manuals can be stored and viewed at the web site and can also be made available for download. The downloaded document has all the formatting, pictures and drawings as the original. You can put an Acrobat document on your network as is or print it and distribute copies. This feature is only common on software and PdM sites not on general equipment sites. For example a manufacturer of Air Conditioner units for aircraft, Keith Products has downloadable versions of the O & M manuals on their web site (pushing the button for manual TR-134) downloads that manual.

KEITH PRODUCTS
Technical Information
provides support to our customers.

R134A Air Conditioning
System Service Manual
Document No. TR-134

R12 Air Conditioning
System Service Manual
Document No. TR-128

Cessna Citation 500/501 Air Conditioning
Maintenance Manual with Illustrated Parts List
Document No. 82-75-010-1SM

Cessna Citation 550/560 Air Conditioning
Maintenance Manual with Illustrated Parts List
Document No. 82-02-010-1SM

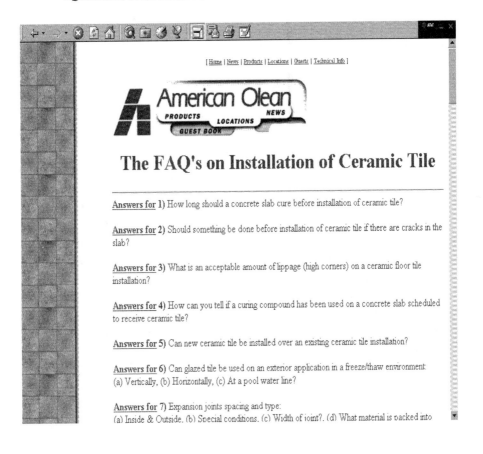

[Home | News | Products | Locations | Guests | Technical Info]

American Olean
PRODUCTS   LOCATIONS   NEWS
GUEST BOOK

The FAQ's on Installation of Ceramic Tile

Answers for 1) How long should a concrete slab cure before installation of ceramic tile?

Answers for 2) Should something be done before installation of ceramic tile if there are cracks in the slab?

Answers for 3) What is an acceptable amount of lippage (high corners) on a ceramic floor tile installation?

Answers for 4) How can you tell if a curing compound has been used on a concrete slab scheduled to receive ceramic tile?

Answers for 5) Can new ceramic tile be installed over an existing ceramic tile installation?

Answers for 6) Can glazed tile be used on an exterior application in a freeze/thaw environment: (a) Vertically, (b) Horizontally, (c) At a pool water line?

Answers for 7) Expansion joints spacing and type: (a) Inside & Outside. (b) Special conditions. (c) Width of joint?. (d) What material is packed into

5. Being able to review FAQ (frequently asked questions) files when installing new equipment or when making major repairs. Some 90% or more of all questions about new equipment have been asked many times before. A focused list with good search capabilities would give the customer with an installation or major repair problem what they need; immediate answers in record time. All information would be available around the clock. The best example of this capability is on one of the major software sites. Your problem has happened before. This capability is common.

6. Access to on-line parts stock list and lead time inquiry. Grainger and McMaster-Carr have taken most of the pain out of purchasing MRO items. Few OEM sites are as sophisticated. One CMMS (Datastream) has a built-in link to Grainger's web site. This is an area ripe for rapid change in the coming 12 months. Making parts ordering easy, quick and painless is essential for maximum profit. Most profits for OEMs come from the parts side of their business. This feature is an uncommon capability.

Purchasing parts for existing machinery is an area of great opportunity for the maintenance field. Dr. Mark Goldstein speaks about the COA (cost of acquisition) of maintenance parts as being poorly understood and, in some cases, orders of magnitude higher more costly than commonly thought. The Internet can slash the COA cost by:

A. speeding the search for the part.

B. streamlining the purchasing process.

C. reducing the time expediting and tracking shipment

D. with EDI (electronic data interchange-electronic billing and payment) reduce bill payment effort.

7. Ability to order spare parts directly on-line. See #6. Companies want to optimize their supply chain. In the new paradigm, our computer can talk to the vendor's ordering system computer and find out accurately when a shipment will take place. We can then accurately schedule PM, PCR (planned component replacement), or any other scheduled activity. This capability will put us a step closer to JITS (just-in-time-storeroom). This is an uncommon feature. For example, you can order anything in the catalog on-line from W.W. Grainger, but you have to go off-line to order spare parts.

8. Ability to check on the status of both parts orders and new equipment delivery status. It is good customer service to provide real time information of this type and it cuts your costs too! One common way to rate vendors is not by the actual lead-time, but by the reliability of their promises. This is an uncommon capability.

9. Lists of approved installers and service companies. Like many of the items on this listing, approved installers and service companies shows that you go the extra yard for the customer. Some of the more innovative OEMs have web links to their distributors and installers (I wonder if they have advanced to the point of charging for the advertising?). This is a very common capability.

10. Maintaining user groups and discussion forums. Discussion forums are common in the magazine world and somewhat less common among the OEMs. If the OEM is a large one, there might be unaffiliated newsgroups run by users. This is not too unusual a capability. A complete discussion of this capability can be seen in Chapter 11.

11. A place to send ideas and operating and maintenance data. A company directory with E-mail links so you can communicate with the right people is surprisingly missing from a majority of sites. When you have important feedback or a great idea, wouldn't it be great to send it directly to the key person herself/himself. Features 11 and 12 are not fully developed (all mail usually goes to a company postmaster who forwards it along). Carrier Corp does better than most. They list all the E-mail addresses in each office (below is the office in Rochester, NY). To send a message, just click on the address and a pre-addressed message form pops up, all ready to go:

Contacts:

Fred Hale, *Area Manager*
fred.hale@carrier.utc.com

Robert Romano, *Territory Service Manager*
robert.j.romano@carrier.utc.com

Sharan Birdee, *Account Executive*
sharan.birdee@carrier.utc.com

Jed Curran, *Account Executive*
jed.curran@carrier.utc.com

Jerry Warren, *Service Account Manager*
jerry.warren@carrier.utc.com

12. Complete addresses of manufacturing plants and local sales offices. This general information should be part of any web site. This is very common (sometimes with a picture of the building!).

13. Useful information, articles, tables and formulas. Where else can we find really specific data or formulas. These features are the public service side of the web site. Space is sufficiently cheap to encourage a little bit

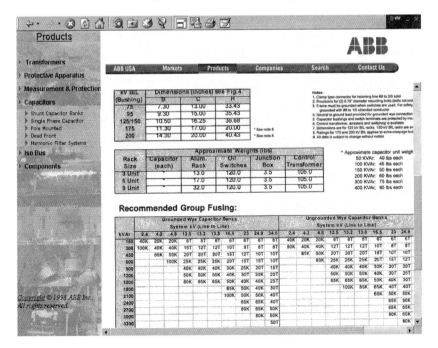

extra. This is a common capability.

14. Links to useful sites in the field. Links to non-competitive sites are always welcome. This is a common capability.

15. Vision for the future: Linkage between maintenance departments and OEM engineering departments. One half of the linkage is on-line access for the maintenance department to OEM engineering data. On-line access for the OEM engineers to maintenance department repair records is the other half. This two-way link would facilitate improvements in OEM product reliability and reduce costs of maintenance. As in other maintenance arenas the OEM-user relationship becomes more of a partnership. Both parties get to contribute their own unique expertise. (This is my daydream for the future: we all can dream, can't we?).

FINDING WHAT YOU NEED

Of the three ways to find information mentioned at the beginning of the chapter, the search engine sites are the most common and most popular way to search for information on the WWW. In the chapter on search engines, we discussed the differences between directory and spider sites. Searching for information on OEM equipment is one area where the difference between the types of search engine makes a difference.

When you seek a major manufacturer's web site or home page, use a directory search engine such as Yahoo. Directory sites return fewer, more directed hits and will lead you to the correct site more easily.

The spider sites are superior for searches for information on specific model numbers of equipment. Use Excite or AltaVista since they cover the whole web (and newsgroups if your request). You enter the exact model number with upper case (where required) and the numbers in the search screen (such as T-131sd). If you are lucky, the hits will be pages within web sites that mention that number. If the model number is widely used, you might also find newsgroup postings that mention your exact asset.

In one search using a model number, we found the OEM's specification page for that model as well as a page from a used equipment dealer with that specific model for sale. When you find manuals, drawings, and specification sheets for your model, bookmark the pages for future reference.

Project: Look up all of the different models of equipment in your plant on the Internet and bookmark the web pages you find. Consider a structure with the model numbers the higher level folder and the type of document a sub-folder.

Folder	Sub-folder
T12323	Maintenance manual
	Sales specifications
	Used equipment dealer
	Discussion group
T2212A	O & M Manual
	Sales specifications
	Parts vendor

All of your model numbers...

SAMPLE SITES TO VISIT

There are tens of thousands of OEM or equipment sites globally. Here are some samples. We recommend that you locate the vendors of your equipment and bookmark those sites as a starting point.

A generalized resource is available from the *New Equipment Digest magazine:* **http://www.newequipment.com.** Penton Publishing has moved its popular New Equipment Digest to the WWW. The site has more than 5000 descriptions of industrial products in various categories. The site is designed around news releases sent to the magazine. The site can fax your request for additional information directly to the appropriate manufacturer. You can also sign-up for an E-mail service, which sends you notices when new products come out.

This site is good for looking up the latest in a specific area. Not good for general information on a topic. It has extensive information for vendors wanting to use the site. It requires complete log-in, which is keyed to your telephone number and has communications to the company.

http://www.ge.com/edc/index.html General Electric.

http://www.herchem.com/ Hercules Chemical company site has links to products of interest to the maintenance department.

http://www.cpmt.com Climax machine tools are portable. They facilitate repairs to even larger tools. This site includes pictures, E-mail, technical documents for download, and a web newsletter.

http://www.trane.com Trane is the largest HVAC company. Some of the highlights of the Trane site include dealer selection, free literature, software analysis tools, jobs, training opportunities, and a library.

http://www.iac.honeywell.com/ Complete information on the Honeywell product line in automation and controls. Thr site includes access to user groups, articles, product information, case studies and technical information.

http://www.johnsoncontrols.com/cg One of the leaders of building controls and automation. They are located in more then 40 countries; Thus, the Internet is a natural home.

http://www.remstar.com Check this site out if you are interested in tool storage and retrieval (with automated charge-out).

http://darex.com/sharpeners. Cutting tool users have access to 24 hour online catalog, FAQ (frequently asked questions)and new product announcements. They specialize in drill and mill sharpeners.

http://www.usinternet.com/catpumps. The pump people really found the World Wide Web.

USED EQUIPMENT

One of the most powerful capabilities the Internet provides is the ability to list items for sale. Classified ad services, for sale sites, dealers, distributors abound on the Internet. A simple search on the machine name (milling machine) and used machine, equipment, or the manufacturers name (Bridgeport) will result in a wide selection of sites.

http://www.stanleyproctor.com A great site for valves, pumps. motors.

http://www.traderonline.com/equip/index.shtml This site is designed for buyers and sellers of heavy equipment. You can see from their web page that there is a lot of buying and selling going on.

http://www.gecsn.com/ GE also has a large used equipment sales operation which disposes of assets from GE Capital and GE Rental operations.

http://machinetools.com/ is a supermarket for buying, selling, financing, and auctioning machine tools over the Internet.

MAINTENANCE DISTRIBUTORS

To maximize effectiveness maintenance catalog sites should follow the rules of good sites recognized by the catalog-marketing professionals. In fact, what we need merges very well with what is recommended by these pros (Target Marketing 11/96 p12). I've taken the liberty to restate their ideas in terms of maintenance distribution issues.

1. The strength of the Internet is in the ability to supply huge amounts of information at relatively low increased cost. In other words, after the site is running, the cost of an extra 10 or 100 pages in nominal. So the rule is, tell the whole story so that the customers can decide for themselves. As maintenance professionals we need to know the sizes, specs, power requirements, in short everything, to make the best decision. Keep each page short and well focused with logical links to additional information.

2. The Internet is not a mass market. The visitors to your site come from every industry, country, and level in their company. The Internet is more like a bunch of small niche markets mixed together.

3. The response to a web site is inversely proportional to its complexity. The easier to use the more it will be used..

4. Free things are good, especially if they help maintenance people do their jobs.

5. Each screen should be able to stand-alone. People frequently get a URL for a sub-page and bypass the home page. Have a button back to the home page, to a site map and a button to communicate with the company on all pages.

6. Be careful of jargon, idioms, and humor because the audience is global. They might be accidentally alienated by what they see or read.

7. Use Be careful of jargon, idioms, and humor because the audience is global. They might be accidentally alienated by what they see or read.

http://www.grainger.com W.W. Grainger Inc. is the distributor of choice for all maintenance departments in factory, building and fleet maintenance worlds. They were first to the web and has the deepest listing of web based

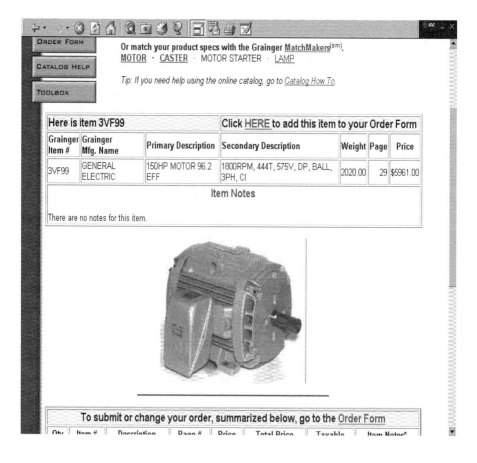

offerings (the tour of the web in chapter 3 stopped at Grainger's site). Site is designed to replace the catalog. 206,000 products are featured. W.W. Grainger's web site raises the bar for smaller industrial distributors. In some fields (such as CDNow in the music industry) the little guy preceded the giants into the industry. In industrial distribution the giants were among the first.

Some of the capabilities include Site map, quote, electronic commerce, and discussion forums. Extensive search for products. Search by type, manufacturer, etc. A search for a 150 HP motor turned up:

Other services include logistics consulting and on-line parts inquiry leading to off-line parts ordering. Pages can be read directly from the newer browsers Has communications to company. Ability to order CD and paper catalog.

http://www.mcmastercarr.com (site was under construction in 7/98) McMaster-Carr is the distributor of choice for heavy industry. They have brought a powerful presence to the web. Site is designed around their catalog. They claim 250,000 MRO items available on-line. Extensive search for products. Search by type, manufacturer, etc. Site map, quote, and electronic commerce. Has a discussion forum. Has communications to company. Ability to order CD and paper catalog Requires Adobe Acrobat to read catalog pages. Pages are downloaded (not directly viewed)

http://www.motion-industries.com Motion Industries are national distributors of power transmission components. They offer a significant value added service from their local stores. We expect nothing less from the web site (and get it).

http://www.industry.net/c/orgindex/ida The home page of the Industrial Distribution Association. This is the association for distributors of MRO items.

CMMS, PREDICTIVE MAINTENANCE AND CONSULTANT SITES

We expect significant presence from CMMS vendors on the Internet and we are not disappointed. Computer software and hardware vendors were one of the first industries to flock to the Internet; the CMMS vendors followed tradition.

There are three types of visitors to a CMMS site. These types of visitors have very different needs. The vendor uses an Internet presence to serve their constituencies cost effectively. CMMS webmasters should plan their sites with almost independent tracks reflecting these three types of visitors.

The first type of visitor is an organization that is shopping for a new system. "Shopping for a system is a daunting undertaking. There are 200 or more vendors of software for maintenance. There are additional 250 vendors in specialized areas such as fleet maintenance, building maintenance, etc" from the book *Managing Factory Maintenance* by Joel Levitt.

For software sites to be useful for these prospects (people shopping for systems), the site should include:

Prospect Needs From a CMM Website
- A concise description of the software/hardware capabilities.
- Tutorial or storyboard that shows some of the capabilities in more detail.
- Downloadable presentations or tutorials.
- Access to manuals.
- Stories from or about users.
- Reprints of stories and reviews from the trade press.
- E-mail to sales staff for quotation and potential site visit appointments.
- E-mail to technical departments for interface questions, system requirements, standards.

☐ Downloadable no-charge working demo.
☐ Ability to purchase software directly off the web.
☐ Searchable FAQ files.
☐ Forum, newsgroup, chat room or other user group mechanism to ask embarrassing questions.
☐ Links to installers, VARs (value added resellers are organizations that can supply the software, hardware, network, training and installation services).

The second type of visitor is trying to find support during the important software implementation phase. The Internet can serve as a landmark when it seems as though the whole project has gone adrift. "The reason that systems fail in their implementation rarely has to do with the lack of reports or inability to get information out but rather from a basic misfit between the system and the existing maintenance culture or organizational requirements. If your culture requires mechanics to keep log books and the system you chose doesn't support that format then they might rebel against the duplicated effort. If you use a 16 digit asset number for accounting purposes then the 10 digit field will not do" also from *Managing Factory Maintenance*. The Internet sites lend a hand with these problems by helping you see a possible route from where you are to where you want to be.

Implementer's Needs are Somewhere Between Prospects and Existing Users

☐ Tutorials that show how to use the system.
☐ PowerPoint presentations to introduce the system and help sell it to all maintenance stakeholders.
☐ E-mail to technical departments for set-up questions, back-up questions and configuration issues.
☐ Searchable FAQ files.
☐ Forum, newsgroup, chat room, or other user group mechanism to ask start-up questions and to talk to people already through the process.
☐ Access to manuals (when they shipped you one, but you need more now).
☐ Success stories from or about users.
☐ Reprints of stories and reviews from the trade press (to send to your boss when needed).

The third type of visitor is the existing user. The web site builds the bond between them and the vendor. If you look at the Microsoft site, you will see most of the resources are designed around supporting existing users. (It certainly doesn't hurt, that almost everyone on the planet is already a Microsoft customer). At microsoft.com it is easy to get upgrades and newsletters. (There are ten or more newsletters for different interest groups). Questions can be researched, and the company offers plenty of free-bees.

Existing Users Have Slightly Different Needs
- ☐ FAQ files need to go well beyond novice questions.
- ☐ Links to knowledgeable insiders (possibly even programmer staff) could save enormous time.
- ☐ Availability of plug-ins from 3rd parts
- ☐ Availability of downloadable patches, upgrades and new versions.
- ☐ Newsletter to find out important changes in the vendor organization (mergers, acquisitions) and any good gossip.
- ☐ Access to manuals (when you lose yours).
- ☐ User groups are essential to trade work-arounds and usage tips that the vendor doesn't even know.
- ☐ The easy access to the latest version and the ability to try out beta-test versions.
- ☐ Power users get to know things about the system that the designer doesn't even know. It pays dividends to the CMMS vendor to stay close to some of these power users.

The ideal CMMS or Predictive maintenance (PdM) site has several attributes. The site should be easy to understand, use and navigate. The home pages need to be simple, easy to understand and quick loading. Given the current state of the web before the (anticipated) speed-up occurs, graphics, sound files and video should be kept to a minimum or made optional to speed viewing time. Links to graphics are okay as long as you warn people about the file sizes.

EXCITEMENT IN THE CMMS FIELD

In their books and seminars the business reengineering gurus Champy and Hammer (*Reengineering the Corporation*, Harper Business, 1994) speak extensively about enabling technology. An enabling technology is a technology that allows a new, efficient, process to emerge that would have been impossible with the old technology. The Internet itself enables a new CMMS or PdM process for larger organizations. We are seeing the very beginning of WWW based CMMS and PdM.

Innovative CMMS vendors are just (at publication date) rolling out a new breed of network CMMS. The new systems allow remote sites to sign on to a company CMMS web site and act as though the CMMS is right in their office. The data entry clerk, mechanic or planner can enter work orders, receive materials, generate reports and make inquires just like a local system. The only difference is that the CMMS is purchased for an Internet server and the screens are browser capable.

The use of the Internet will allow centralized data warehouses. These warehouses will facilitate sophisticated analysis that is currently missing from most implementations (because the statistical population is too small).

The only comparable analysis situation is the FAA computer in Oklahoma City. All domestic airlines enter maintenance records for their aircraft. The advantage is that the maintenance statistics of all the components of a certain model are aggregated into one model. This allows a large enough universe for some heavy-duty analysis. The Internet will allow the GSA, for example, analyze all of their Cleaver-Brooks boilers or Trane air handlers.

Using a centralized system via the Internet would facilitate communication within the plant engineering and maintenance department's worldwide. It would also allow standards for repairs, coding, PM tasks, part use, which heretofore has been impossible to achieve.

"The future might be more similar to the past then anyone's wants to admit," says Dr. Mark Goldstein. He was speaking about the concept of metered usage of CMMS or PdM software via the Internet. The last time this model was in the vogue was in the time-share days. In the 1970s and 1980s it was common to rent computer time on a time-share basis. The vendors were either professional time-share companies or large organizations with unused capacity. In some fee structures you paid for CPU seconds, connect time, storage bytes and the applications you used.

In the future, he postulates, you might sign on to a PSDI web site, enter your companies' password and rent Maximo, the processor and some disk space. That would be very interesting to both CMMS vendors (steady revenue stream, access to real data to test new components) and users (less outlay, fewer people, automatic upgrades, international access, help desk are experts).

EVALUATING CMM PACKAGES

It seems strange but you can use the Internet to find the answers to some of the questions about how to shop for a CMMS (can be found online at http://www.asd-info.com/eval.htm)

40 Ease-of-Use Criteria for Evaluating CMMS & FMS Packages

Evaluating Computerized Maintenance Management Systems (CMMS & FMS) is difficult at best. Most of the leading packages offer the important modules that you are seeking (Work Orders, Service Requests, Preventive Maintenance, Equipment History, Inventory Control, Purchasing Functions and Reporting); all claim to be easy-to-use. It is very easy to fall into a method that involves comparing the look and completeness of each system on a screen by screen basis; this can lead to disappointment after purchase.

Surveys of thousands of Maintenance Management Software users have indicated that you will benefit by ensuring that your evaluations consider CMMS & FMS products under conditions that will reflect the actual ways in which your staff will work with the product each day as they perform their duties. This clearly is not a mode in which the user moves in an orderly fashion through a menu structure from screen to screen. Instead he or she will be on one screen and need to quickly get something from other screens without a series of steps and screens to get to that point and return, i.e. to move across modules rather than up/down/across menu structures. You will be well served to structure your evaluations to reflect these requirements and to observe the steps required to accomplish them. Suggestions follow; the CMMS system ought to be able to accomplish these activities with 2 or 3 steps, not 5, 8, 10 or more.

From the Equipment Screen
- ☐ Create a work order for this equipment.
- ☐ Check the maintenance schedule for this equipment.
- ☐ Find all open work orders for this equipment
- ☐ Find all history work orders for this equipment or all work orders this year or last year.
- ☐ Find all maintenance procedures or services that apply for this equipment.
- ☐ Create a service request for this equipment.
- ☐ Find all open service requests for this equipment.
- ☐ Find all historical service requests for this equipment or all for this year or all for last year.
- ☐ Look at history for similar equipment to find work orders dealing with a similar problem.
- ☐ Find all equipment that will not be usable if this equipment is removed from service or if main power is interrupted for this equipment.
- ☐ Remove this equipment from service for an extended period of time and automatically suspend PM schedules.
- ☐ View and/or print attachments that apply to this equipment; create a new attachment; link something from outside the CMMS as an attachment.

From the Location Screen
- ☐ Create a work order for this location.
- ☐ Check the maintenance schedule for this location.
- ☐ Find all open work orders for this location.
- ☐ Find all history work orders for this location or all work orders this year or last year.

□ Find all maintenance procedures or services that apply for this location.

□ Create a service request for this location.

□ Find all open service requests for this location.

□ Find all historical service requests for this location or all for this year or all for last year.

□ Look at history for similar locations to find work orders dealing with a similar problem.

□ Find all locations that will not be usable if this location is removed from service or if main power is interrupted for this location.

□ View and/or print attachments that apply to this location; create a new attachment; link something from outside the CMMS as an attachment.

From the Work Screen

□ Find all open work orders for the same equipment or location.

□ Find all history work orders for the same equipment or location.

□ Find all work orders for this year or last year for the same equipment or location.

□ Check history of this work order.

□ Import a pre-existing maintenance procedure for this equipment/ location.

□ Convert this work order to a multi-task work order with an additional task requiring a different work group, trade, outside contractor, subassembly or cost account.

□ Record actual labor for the work order without closing the work order so t that work in progress can be tracked and reported.

□ Record labor hours for a craftsman with multiple entries for multiple days.

□ Add parts specific to a single task of the work order.

□ Find all Service Requests linked to this work order; open any of them.

□ Automatically close a service request when this work order completes if no other work orders are still open for the service request.

□ View and/or print attachments pertaining to the work order, to the equipment, to the location, to the staff assigned, to the contractor assigned, to the parts or tools used.

□ Add new equipment, locations, staff, contractors, parts, accounts or others.

□ Check maintenance schedules for the equipment or location this work order is servicing.

□ Check all stocking locations for a part required, determine if a part is on order.

□ Order a part (through internal parts ordering staff or to outside vendor) via any of a variety of methods (direct, email, fax, and modem).

From a Service Request Screen

□ Create a work order.

□ Find duplicate service requests.

- ☐ Find all open service requests and linked work orders for the same equipment or location.
- ☐ Find historical service requests and linked work orders for the same equipment or location - all, last year or this year.
- ☐ Change status of the service request - reschedule, send for approval, can cel, other.
- ☐ Link existing work orders to the service request.
- ☐ Allow the service request originator to electronically monitor status of the service request.
- ☐ Fax, email, modem the service request and/or linked work orders.

These are intended as merely a sampling of the activities that the maintenance software should handle with ease and with a minimum number of mouse clicks and/or screen changes. These suggestions are provided by Advanced Software Designs after reviewing surveys covering thousands of CMMS & FMS users over the past 5 years

Most of the CMMS vendors have web sites. Their web sites vary from automated brochures to an approximation of the ideal site with many of the sections listed above.

A complete list of CMMS vendors can be found at the International Industrial Engineering Association site (this one is very complete): http://www.iinet.net.au/~sdunn/maintenance/CMMS_vendors.html

A commercial CMMS supersite that lists many packages (including some unknown ones) is the TurboGuide site. TurboGuide lists software products belonging to manufacturing and equipment maintenance. The following is an excerpted list of links. http://www.turboguide.com/

http://www.benchmate.com Benchmate Systems is a basic maintenance system. The site is simple and quick with a downloadable demo (750k). Modules shown include equipment, service order, PM, inventory. Ability to communicate by E-mail. No pricing, quotes, or even company address. No support for implementation or existing users. No external links.

http://burkesystems.com/Software.htm Burke & Associates has a different slant on the CMMS field. They call themselves decision support systems specialists. The clients they list are all municipalities. The system grew from a service order system to its full implementation today. The site has a very sketchy description of the system with no examples of screens, reports or inquiries. Site includes a user group (protected by sign-in requirements). Little support for people choosing a system, but some support for owners through the forum. No external links.

Manufacturing Equipment Maintenance	
Software Title	Software Platform
Advanced Maintenance Management System, The (AMMS)	MS-DOS
BCSCRIB	MS-DOS
COMPASS	MVS; VM; VMS; VSE; Windows 3.x; Windows
CRIB MASTER	MS-DOS
Compufix Maintenance Management System (CPX/MM)	MS-DOS; MVS; VMS; VSE
Computerized Maint Management System Plus	OS/400; MVS; VMS; VSE
Confined Space Entry Tracking Module	MS-DOS; VMS; SUN OS
Diagnostic Advisor	Windows 3.x; MAC System 6; MAC System 7
Expert Maintenance Management	Novell NetWare; Windows 3.x; Windows 95;
Expert Tool/Gage Management	Novell NetWare; Windows 3.x; Windows 95;
Facilities and Resources	MS-DOS; Windows 3.x; Windows NT; Novell
GAGE MASTER	MS-DOS
Hemisphere	MS-DOS; Windows 3.x; OS/2; VMS; SCO UNIX
IQs PREVENT(TM) for	PM DEC UNIX; MS-DOS; Novell NetWare; OS/2;
Life Cycle Cost Analysis — LC2M	MS-DOS
MINIMONITOR	MS-DOS
MOTORMONITOR	MS-DOS
Main/Tracker	OS/400
MainPlan/EQ	MAC System 6; MAC System 7
Maintenance	Pro MAC System 6; MAC System 7; Windows 3.x;
Proteus CMMS (Equip maint help desk)	A/UX; AIX; DEC UNIX; HP-UX; Novell
Rainbow Bar Code Data Collection Kits	MS-DOS; VMS; UNIX; SUN OS
Reliability (RELIAB)	MS-DOS
SMART/CMS Calibration Manage Sys	MS-DOS; Windows 3.x; MVS; Solaris; VMS
SMART/MMS Mainten Manage System	MS-DOS; Windows 3.x; MVS; Solaris; VMS;
SMART/SCSS Stock Catalog Search/Support Sys	MS-DOS; Win3.x; MVS; Solaris; VMS;
SMART/SIMS Spare Inven Manage Sys	MS-DOS; Win 3.x; MVS; Solaris; VMS;
SMIDS — Space Modeling and Interference Detection System	VMS; SUN OS
Test Bench	MS-DOS; OS/2; VMS; SUN OS
UPTIME	MS-DOS; Windows 3.x; OS/2
Vessel Inspection Manager	MS-DOS

http://cogz.com/cmmsa.htm Advanced Maintenance Solutions. Two packages are promoted (copied from the site):

COGZ EZ   focuses on Preventative Maintenance and Work Orders. Supporting features such as inventory and purchasing are also included. COGZ EZ can be easily upgraded to COGZ at any time.

COGZ   includes additional reports and data analysis features.

The main sales pitch is ease of use and ease of set up. Downloadable demo on site. Links to news releases. Upgrades available for registered users on-line. E-mail support available. No external links.

http://mainboss.com Desktop Innovations is the company behind the Mainboss maintenance software for Windows(r). It is a Canadian program with a unique approach. The downloads are available in French, Spanish and English:

Pour la version francáise, veuillez faire un click ici...

Para la version en Espanol, click aqui...

This site is setup in a logical way. It clearly describes the product and gives explicit pricing for the package and for the annual service agreement. The support screen for existing users shows a FAQ database, on-line support, and updates. A unique feature for implementation is access to a sample database. This allows experimentation and in-depth investigation of how to handle certain issues.

In order to see samples of the system itself you would have to download the demo since there are no examples from the system on the site.

http://www.dstm.com Datastream is the market leader in terms of units of sales. (With some recent acquisitions, they are probably also approaching sales volume leadership). Their home page is graphics intensive with easily understood links. The site lists seminars, describes the company, and summarizes the product, support, partners, etc.

The support screen looks as follows:

The Datastream Technical support staff is trained to answer your questions about our software. As a Technical Support subscriber, you are entitled to all Datastream support services including telephone support using our toll-free support number, access to our bulletin board, Internet email support, and free updates to our products.

To receive these services, you must be a TechSupport subscriber. If you're not a subscriber, but would like to be, call TechSupport Renewals at 1-800-955-6775. You can renew your yearly TechSupport subscription by using our registration form provided on the web for your convenience. Inside the U.S. Call:

Support 1 (800)365-6775

Support Fax 1 (864)422-5233

Support Renewal Fax 1 (864)422-5321

If you are already a subscriber, you can email us at support@dstm.com. If you would like to call us with your support needs, a list of phone numbers is provided here.

The FAQ database search engine has a strange idea of what is related to what. A query on PM gave me information on how to load the zebra printer, how to get a report to print right (set the date to the beginning of the week), and some unrelated system errors. Otherwise a useful function. No outside links. No downloads available.

http://www.desmaint.com/ The site of DMSI. The organization could also be in the consulting part of the chapter. In addition to CMMS, it sells a lubrication management system and a condition monitoring system. This site seemed to take a long time to load (graphic intensive). Brochures are available in .pdf format. Demos for all programs are available.

http://www.dpsi-cmms.com/ Graphic intensive site. Seemed to be on a fast server or well designed to load fairly quickly. This site is well organized and complete in the areas it covers. Verbal descriptions of modules with no examples, screen shots, or reports. Site has capabilities for downloading both demo software and updates to existing users. Users are given contact E-mail addresses directly in several departments such as training, consulting, and technical support. Complete site map. Search engine for site was not too helpful. Site has a good section of testimonials from existing customers. Has the best link site of all the CMMS vendors. The link site is worth the visit itself.

http://www.eagleone.com Eagle Technology maintenance management software has something unique. They allow you to download a simple basic system for free. For many small departments this might be the entry ticket. The site can be viewed in English, German and Spanish. It gives a list of reasons why one should computerize which could be useful when trying to justify a system. When a properly applied program of automated maintenance is put into place:

- Staff and equipment operate more efficiently and cost-effectively
- Equipment failures can be anticipated and prevented
- Workloads can be balanced to correspond to available manpower
- Paperwork and communication costs are greatly reduced
- Equipment and parts costs are optimized
- Reduces unnecessary failures, and downtime
- Increases equipment utilization
- Lowers inventory levels, carrying costs
- Reduces paperwork
- Easily track maintenance cost history
- Improves overall productivity
- Loose creditability with customers
- All information is easily accessible
- Improves customer satisfaction

Site has a pretty complete chart comparing its various products. Support section protected by password for registered users only.

http://www.jbsystems.com JB Systems, founded in 1983, is one of the old-est organizations in the maintenance software business. They have offices worldwide. The thing that strikes me about the Mainsaver site is the variety of ways they support users. Because of the longevity, one of the unique attributes on the site is an entire page of local user groups with people's names and their contact phone numbers. It's quite impressive to see twenty different user groups. In addition to the user groups there is an annual users conference (designed by a user committee). Preceding the conference is a week of training workshops.

Extensive classes are available in the Los Angeles area. The JB Systems newsletter comes out quarterly. JB Systems maintains an FTP site for patches, upgrades and plug-ins. In addition to the support mentioned they have FAQ files, faxback services, and E-mail support as well as phone.

The package integrates with most of the enterprise computing systems including SAP and People Soft. The product introduction is all text with photographs showing people using the system. No screen shots or reports are available. They showed web capabilities (remote system access and update through the Internet).

http://www.avantis.marcam.com Avantis is a large scale CMMS where installations can cost an average of $400,000. The site is solid and easy to navigate. The product descriptions are sketchy and there is no direct download capability (you can order a free CD demo off the site). You can clearly see that the site is not designed to sell systems. It is a limited web presence for existing users to link into the forums and BBS and for casual surfers to generate inquiries for the sales force to handle.

http://www.microwst.com Microwest AMMS (Advanced Maintenance Management System). The software may be great but the web site looked like an afternoon project. There were no details about any aspect of the system, services, support or company itself. I assume they just got on-line. Hopefully they will read the articles that say that the Internet lends itself to details and the more the better. There was a demo registration (for snail mail delivery) but no downloadable demo. No FAQ files easily accessible. No site map beyond contents bar (and none needed).

http://www.norwichtech.com Norwich Technologies, Maintenance Master has some unique capabilities in the area of collecting information from machines and generating work orders based on the conditions reported. The site wasn't specific enough to see where they were really going with that capability. The site seemed to have more graphics than would be necessary to tell the story (in other words the graphics didn't add to the story telling). There was easy access to E-mail and there was a downloadable demo. No FAQ files easily accessible. No site map beyond contents bar (none needed).

http://www.omni-comp.com Omnicomp is a product of Service Call Express, which is a subsidiary of Enron Corp. Enron (a $15 billion energy company). Omnicom is a 20-year-old subsidiary that started in the energy management side of the business. In the description of the product, more space was given to hardware configuration then to software capabilities. I wonder why. In their contact section they even include a map to get to their facility and a list of hotels with hot links! No usable resources for existing users or new users in implementation. No FAQ files easily accessible.

http://www.prismcc.com/famis Prism Computer, FAMIS system. Pretty complete description of the system (which seems to be a part of a family of applications for business). Support section protected by a password. The modules of the maintenance management were listed with links to short descriptions of the modules:

INTRODUCTION	PURCHASING MODULE	REPORTS
SERVICE REQUESTS	INTEGRATION	EMPLOYEE MODULE
WORK ORDER CONTROL MODULE	EQUIPMENT MODULE	LOCATION MODULE
JOB COSTING MODULE	ATTACHMENT FEATURES	
PREVENTIVE MAINTENANCE MODULE	PROJECT MANAGEMENT MODULE	
MISCELLANEOUS FEATURES	PLANNING & ESTIMATING FEATURES	

One of the most interesting things was their statement about the Y2K bug:

Year 2000 Statement

FAMIS 5.1 and earlier releases, in addition to early FAMIS 6.0 beta releases are NOT fully year 2000 compliant. The production release of FAMIS 6.0 will be year 2000 compliant. It will satisfy five main factors with respect to date datatype processing:

1.Correctly handle date information before, during, and after January 1, 2000 accepting date input, providing date output and performing calculation on dates or portions of dates.

2.Function according to the documentation before, during and after January 1, 2000 without changes in operation resulting from the advent of the new century assuming correct configuration.

3.Where appropriate, respond to two-digit date input in a way that resolves the ambiguity as to century in a disclosed, defined and pre-determined manner.

4.Store and provide output of date information in ways that are unambiguous as to century.

Manage the leap year occurring in the year 2000, following the quad-centennial rule.

Site has a good event section and many users can look forward to the FAMIS conference in Disneyland! The site had one flaw; a message of a bug in the site code kept coming up (running Explorer 4.0). That did not inspire confidence. It is easy to contact them through the spinning E-mail button left in the frame around the site. The customer list was impressive.

http://www.psdi.com PSDI is one of the recognized leaders in the CMMS field. Their site reflects their position in the industry. It has depth and useful capabilities for all three types of users. Among the most interesting sections are the case studies about successful installations. There is a separate user group web site protected by a password. The descriptions of the system are detailed and include screen shots and written explanations. One clever idea is the use of the Internet for a toll free phone call with a PSDI representative. PSDI is one of the few companies where a VAR or reseller (in this case Projetech) might organize the user group:

Midwest MAXIMO(r) Users' Group Meeting
October 19, 1998
Radisson Inn - Cincinnati Airport
Note: This will be the only meeting for MAXIMO(r)
Users sponsored by Projetech in 1998. Mark the date & make plans to join us.
Check back to this page for more details!
For details on any upcoming events,
contact us at(513) 481-4900 or send inquiries to
Beth Farr.Midwest MAXIMO(r) Users' Group Meeting

http://www.walker.com Revere, Inc. was purchased by Walker, Inc. (an enterprise systems house). This is primarily a large organization solution. The web site is dominated by verbal descriptions of each module or capability. Much detail, but no examples; a listing of prestigious customers, but no details. No user capabilities, no FAQ files, no demo (downloadable or otherwise), and no pricing. It is difficult to see how to contact the company from the web site.

http://www.somax.com Somax (advertised as the customized CMMS) starts out the site with real reasons why to computerize. It asks the critical question and then answers its own question.

Who wouldn't want that? The site really starts on the second page (you are left to figure out how to get there). Demo download available. Complete contact information on the 'About Somax' page. One useful function is the ability to download a fully formatted and illustrated catalog in .pdf format. Password protected area for user updates. Three different training

---

SOMAX Information Management Systems

Keeping your maintenance program on track takes more than just a pencil and clipboard. It takes a real information management system designed specifically to support your complex needs.

A system that links your asset management, maintenance, inventory and purchasing functions for more efficiency.

A system that gives you quick access to vital information for better decision-making.

A system that puts an end to lost parts, lost time and lost productivity.

A system that increases your equipment availability.

---

classes already scheduled. No external FAQ files, no customer stories, no system examples, screen shots or reports.

http://www.sofwave.com Sofwave Maintenance Information Systems has a unique option called Pulse that reviews the database and provides analysis. The analysis is action oriented (to do lists of overdue items, etc.), with charts showing what is happening. Easy contact through both E-mail and complete company information. As you can see below, there are several advanced features.

Features

Custom Reporting	EDI Capabilit	Purchasing
Work Order System	Analysis Capabilities	touch-tone
Calibration	Multi-Media	Project Management
Bar Coding	Inventory Management	Predictive Management
Troubleshooting	PM Scheduling	Compliance
OSHA Accessibility	Audit Trail	Conflict Manager
Tool Management	Facility Maintenance	

This is a sales-only site with no capability to support existing users. The technical support section is password protected. No external FAQ files, no customer stories, no system examples screen shots or reports.

http://www.tmasys.com TMA maintenance information systems. One of the few systems that have a section where you can review a variety of reports. They have a link built to the RS Means standards database to facilitate estimating. Complete, albeit unusual, site map with a graphical and an outline mode. One of the best marketing ideas is the competitive upgrade rebate.

Another unique and useful aspect of the TMA site was a complete 15-screen software specification. You can really see what you are getting! They define three different support plans (platinum, gold, silver). They have

> **Dissatisfied with your current CMMS performance and features?**
> **Not getting top-of-the-line client support that you're entitled to?**
>
> ## TMA offers you a way out.
>
> If you're a current owner of any CMMS software version of Datastream(r), Chief/Maximo(r), ACT, DPSITM, or any other TMA competitor's CMMS software, you're entitled to a competitive upgrade rebate on TMA's PowerBaseTM, AFMTM, or AFM-SQLTM software.
> To take advantage of this offer, call TMA sales at 1-800-862-1130 for further details.

both phone and E-mail support. There are pages of clients, especially in the education and healthcare fields. One quibble is that the section was titled user profiles and there were just lists of names. The very complete site had no FAQ files and no forums.

http://www.tswi.com TSW International. There is a complete written description of the system. Good justification for each module (what it would do for me). No examples or stories from users. They maintain a special network for supporting all of their products called CareNet. They offer the following capabilities on CareNet.

- log questions or problems directly to customer service
- search our knowledge base of known solutions and answers
- monitor product enhancements
- keep apprised of product maintenance activities
- download product updates

http://www.westernsoftware.com/products.htm is the home of the Guardian Maintenance Management System. This site is a few pages about the package, objectives and instructions to contact the company. It is a web presence and nothing else.

http://www.aa.net/~winter WinterCress Has a good section they call 'White papers' that explains the basics of a topic such as bar coding. Product descriptions are skeletal with no examples. Site has no client stories, report examples, or details of any type. They include newsletters from 1995? Product support is through phone support (not web).

EXAMPLE OF OTHER KINDS OF SOFTWARE

The maintenance field can use many different types of software to get the job done. One of the things that you notice at sites like the Turbo software is the range of packages that might be of interest to a typical maintenance professional.

RetrieverSoft
Corporation Office: Toll free: 800-242-5656 Fax: 215-424-4284
902 Oak Lane Ave. Philadelphia, PA 19126.

Provides the power to manage

We provide software, services and procedures that turn's piles of information, notes, sketches, documents, booklets into a competitive advantage. With Retriever your information is worth more because it is available when you want it, where you want it, how you want it, and to whom you want it to go.

The cost is very reasonable. System resources are also reasonable. The system has been optimized for several industries including equipment, maintenance, field service, fleet management, building management, operations management, chemical MSDS and new ones are being all the time.

Interested in hearing from a real RetrieverSoft person?

Mail systemsales@retrieverSoft.com

Comments to management: webmaster@retrieverSoft.com

OLD WAY before Retriever

Piles of information consisting of 1121 files, 235 drawings, 197 hand notes in 7 piles, 131 manuals. All are located in 4 file cabinets, 3 bookshelves, and every horizontal surface. Average search time for a specific item 24 minutes. Average filing time 1 minute each item. Average number of items falling through the cracks 2 per week. Average search time when you are not there 2 hours 15 minutes

Issues

No ability to make archive copies of critical documents, drawings, manuals. Personnel files not available to everyone. Impossible to locate small pieces of paper. Cost and time to track down stuff borrowed and not returned is high. More projects and new machines take up more space

NEW WAY WITH RETRIEVER

Everything on computer. Average search time for a specific item 2 minutes. Average filing time 1-10 minutes each item. Average number of items falling through the cracks 2 per year. Average search time when you are not there 5 minutes.

Benefits to you the retriever user

Ability to make archive copies of critical documents, drawings, manuals, total back-up 1 hour. Files available to everyone on network or with access to your computer. Small pieces of paper are as easy to find as large ones. Stuff can be viewed on screen, printed, faxed from system no need to return.

More projects and new machines in same space (huge projects add disk drive)

PREDICTIVE MAINTENANCE PDM SITES

"The ideal situation in maintenance is to be able to peer inside your components and replace them right before they fail. Technology has been improving significantly in this area. Tools are available that can predict corrosion failure on a transformer; examine and videotape boiler tubes, or detect a bearing failure weeks before it happens.

The way predictive maintenance improves reliability is to detect deterioration earlier then it could be detected by manual means. This earlier detection gives the maintenance people more time to intervene. With the longer lead time there is less likelihood of an unscheduled event catching you unaware" from the *Handbook of Maintenance Management* by Joel Levitt.

Web sites for predictive technologies are similar to the sites for CMMS. The companies tend to be a little larger since they are mostly manufacturers. The criterion for excellence is almost identical. One would expect more 'how to' stories since the technology can be used in many applications.

http://academy-of-infrared.net The Academy of Infrared Thermography qualifies as an association, training center and a resource for people wanting to get qualifications in thermography. Site includes a store (with software, books and E-commerce), some association links, and a few industry links. They have three levels of courses and specific application courses. A good place for engineering-oriented infrared people. A partial list of papers available on the site:

> Infrared Thermography Yesterday Today and Tomorrow
> Motor Circuit Analysis for Proactive Maintenance - *Rob Yontz*
> Roof Thermography and FPA Technology - *Bob Miljure*
> Definition of Modern FPA Terms - *Andrew Teich*
> Viewing Main Power Distribution Components - *Robert Andrulis*
> Using Infrared and Ultrasound - *Markus Blaszak*
> Application of Infrared Thermography - *Brian Holmes*
> Thermographic Inspection Of Paper Machines - *Randy Greenall*
> Detect Problems with Cable Terminators, Potheads
>   and Cables - *Ralph Herzog*
> Infrared Thermography at Intel's Fab 8 - *Yaacov Shuval*
> Business Opportunities for IPdM - *Ron Smith*
> Techniques for Reducing Oil Spill Incidents - *Eugene Onyeka*
> Increased Inspection Proficiency and Reduced Report Writing -
>   *David Stonehouse*

http://dingos.com Dingo Maintenance Systems, oil analysis software. This organization is dedicated to bringing the science of oil analysis to the maintenance world. According to their web site:

> *Designed by maintenance professionals*
> *for maintenance professionals.*

Intelligent product features help users take advantage of all that modern oil analysis technology has to offer. Data handling is reduced and

interpretation is made easier, allowing maintenance personnel to:
Focus on maintenance, not software
Prevent expensive equipment failures
Reduce routine maintenance costs
Maximize equipment availability

Dingo's site is under construction and has some interesting capabilities in the pipeline such as an oil analysis forum and a web-enabled service for global organizations to track their oil deterioration.

http://www.entek.com The Entek acquired IRD Corp. in 1996. IRD, founded in 1952, to some degree, invented commercial vibration analysis. The two companies formed a powerhouse in the predictive maintenance business. They have an extensive training calendar with descriptions posted on the site. Education and support (as well as good hardware and software) determine the success of a predictive vendor. Their site is slanted this way, as this example from service section shows

The mailing list is like a newsletter. There are three different lists:

TIPS AND TECHNIQUES

Welcome to the Tips and Techniques Page. Use this online guide to maximize the productivity of your Entek IRD products. Check back here frequently for the latest in product tips and application techniques.

- **Search** the Support Site for Keywords
- **Browse** for information by Topic or Category
- Participate in an ongoing **Discussion**
- Join an Entek IRD **Mailing List**

---

general support, downloads/patches, and bug fixes. The link page is pretty thin with less then a half of a dozen links.

http://www.framatech.com The predictive section of Framatome Technologies can be found at http://www.framatech.com/marketing/Empath.asp and is a very small part of a larger service company for the electric utility industry. They have technology and engineering services to inspect large motors and detect:

EMPATH indicates
- Rotor bar deterioration
- Rotor/stator eccentricity
- Stator phase imbalance

    ☐ Motor speed and slip frequency
    ☐ Gear and belt imperfections
    ☐ Average running current; an indicator of motor torque
    ☐ Stroke time on assemblies with defined start and stop points
    ☐ Changing friction forces
    ☐ Torsional vibration and dynamic loading
    ☐ Detection of bearing defect frequencies

There is contact information, a downloadable .pdf brochure, and many links to power utilities and associations for utility surfers.

http://www.inframetrics.com Inframetrics infrared specialists has an up-to-date site with all the bells and whistles, even music on the home page. You have to have a current browser to see everything. The usual issue with these fancy sites is "where's the beef". This site is also a winner on content too. In the applications section, there are 15 different applications with visible and infrared pictures. The section is very informative. They have detailed product descriptions, but no pricing or ability for E-commerce.

http://www.omega.com The OMEGA site was visited in Chapter 3, the tour of the Internet. It is a powerful site. It has a comprehensive search engine and a complete site map. One unique feature is an E-mail directory. There is a great deal of technical data on each product (like the catalog). You can go on an armchair tour of their company sites in different countries like the Czech Republic, Benelux, Mexico, Latin America, UK and Germany.

http://www.pdma.com PdMA test equipment for motor circuit evaluation and a group in oil analysis. PdMA will evaluate your current predictive maintenance program. They have a simple site that is easy to navigate.

http://reliability.com Reliability is a training and consulting organization that also sells PdM software. Proact is their software tool ($595) for root cause failure analysis. A demo and a .pdf brochure are available for download. Site includes a discussion group. Their site includes many useful links.

http://www.redlake.com/imaging Redlake Camera high-speed video and motion analysis site. Many of the technologies under the predictive banner are not well understood by the general maintenance population. High-speed video is no exception. The site is more of an educational one than a hard sell, E-commerce one. You can tell that because the first item on the home page is a link to FAQ's, examples, jump sites and the second item is the sales literature. Full specifications and no pricing. Communications with company available. No forums, technical support available through the site.

http://www.isigroup.com ISI Group, Inc. There was limited information on either their infrared gun or their software, and no pricing. On the sales side an excellent idea was rentals and refurbished equipment:

To rent	THERMAL IMAGERS FOR RENT Monthly	Weekly
	$2,100.00	$945.00 (week one)
		$630.00 (week two)
		$575.00 (week three)
		$525.00 (week four)

If you purchase a new thermal imager within the rental period, we will apply the rental cost to the new purchase. For descriptions of rentals and rental agreements, please contact us at (800) 821-3642.

Refurbished Equipment
From time to time, I.S.I. Group, Inc., has had an inventory of refurbished VideoTherm(r) cameras available. Please contact us for availability and quotations.

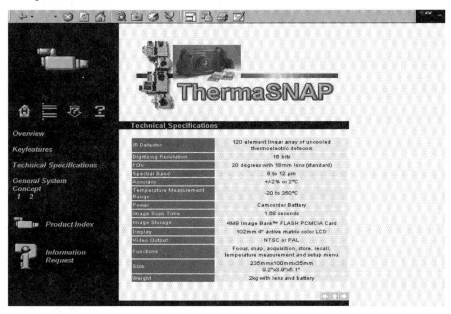

Information about infrared was lacking. Training classes or books were lacking. Very few links on their jumpsite button. The site has several forms to contact the company (including technical and sales support).

http://www.uesystems.com UE Systems offers (via its web site) a full range of modes for preventive maintenance. You can find the following:

> **UE System Inc.** is the worldwide leader in Airborne Ultrasound since 1973. Portable & continuous ultrasonic testing instruments for leak detection, mechanical analysis, and electrical inspection.
>
> **UE Service Partners Inc.** is an affiliation of proven, competent, highly qualified inspectors. Inspection services for mechanical trending, electrical inspection, steam trap, valve and generic leak detection.
>
> **UE Training Systems Inc.** is the world's first certifiable training course for Airborne Ultrasound. Level I, Level II, Level III conforms to ASNT standards recommended practice SNT-TC-1A.
>
> **EPD Technology Corp.** offers inspection, testing, monitoring instruments of the highest quality. Products range from leak detectors and monitors to infrared thermometers, boroscopes, eddy current and metal detectors.
>
> **Instrument Support Services** provides sales and support for all of the UE Group. All sales consultants provide professional, personalized, service. Their extensive training and product knowledge makes them a valuable resource for all their clients.

One of the, all-time best ways to get potential customers to respond is to offer something for free. They have a free gift (I won't give it away and say what it is) in exchange for you filling out a survey. They have an excellent section called detective maintenance that lists a half-dozen mysteries. The stories are written like Sherlock Homes tales, complete with Dr. Watson. The solutions will amaze you. The technical overview is excellent. There is a forum sponsored on the UE technologies. There are several videotapes, training courses, and software (with free demo downloads).

Condition-based maintenance

http://www.libertytech.com Liberty Technologies provides machine condition monitoring. While condition monitoring is not usually part of maintenance, the attributes monitored can provide a useful view of the ongoing condition of the asset. Many of these techniques provide inspection intervals quick enough (1-10ms) to detect and alarm before impending failure. Their site claims:

By combining advanced diagnostic technologies with comprehensive software analysis, Liberty enables plant operators in varying industries to prevent unplanned downtime, improve asset utilization and increase plant safety. Liberty's proprietary products monitor condition and performance by gathering and interpreting operating data of valves, turbines, engines, compressors, motors, motor-driven equipment, pumps and fans. The Company's digital radiography technology improves upon conventional methods in speed, safety and cost. Liberty's products utilize sensors, instruments and proprietary software that capture and log data for trending and analysis.

Alignment

Alignment is not formally like CMMS or predictive maintenance. It is primarily an installation tool. Sophisticated alignment can also be used in an existing plant on existing equipment to lengthen life and reduce energy usage. Since most of the technologies use computers controlling the lasers, the companies tend to be similar to the PdM companies.

http://www.petersontools.com Peterson Alignment Tools, precision alignment tools. It took a few seconds to find my way into the site (it looked like a one-page site until I kept trying). They define alignment as follows:

Shaft alignment (within the context of this Website) refers to the alignment of two shafts joined together by a coupling assembly. As an example, this equipment could be a motor driving a pump circulating fluid (oil, water, etc.) throughout a manufacturing plant.

With few exceptions, motor and pump shafts joined together with a coupling need to be aligned properly, initially when the unit or units are installed, and periodically during the life of the equipment within the plant.

You may ask, "What, exactly, are we aligning with the alignment tool?" The answer: the two shafts that come together when joining a motor and pump assembly never actually touch each other, although they come close. The "coupling" that mates these two shafts together is there to transfer the power produced by the "driver" unit (a motor) to the "driven" unit. The coupling (in general) slides over each shaft and a slotted key and set screw keeps it from spinning on the shafts. Between them in the center, a rubber "spider" fitting makes the needed connection for transfer of power.

The alignment tool (either our model #20RA or #30RA) is then used to perform the alignment. The first step is to mount either kit on the driven side (usually the pump side, also the side that is used for our "0" reference). By following the detailed instructions provided, readings obtained during the procedure will tell you the amount of misalignment present in the form of "shim" amounts needed to be added or taken out under the front and/or back feet of the driver (moveable) unit. Also, any adjustments from side to side (away from or towards you) are provided.

Still don't understand it? Call me, Chris Bowen, right now at (800) 254-4611 and I will be glad to go over it with you.

They have a special download of their complete manual package available in .pdf format. Upgrades and pricing available on the site. Product specifications included pricing. No E-commerce available. Actual discussion of alignment with appropriate math was a useful overview.

http://www.pruftechnik.com Pruftechnik shaft alignment and machine monitoring. Alignment can improve uptime, reduce energy use, and increase equipment life. This company has a worldwide presence (the sites are listed) and a worldwide reputation (200 patents). Their technical documents section has over 35 documents devoted to alignment alone (require .pdf reader). Training courses are available from local centers and are not on the web site. E-mail to company is available. No E-commerce, no prices, no delivery information are available on the site.

http://www.vibralign.com VibrAlign, precision alignment instruments, has a very straightforward and easy-to-use site. They had a useful evaluation grid for laser alignment:

Key Buying Criteria	Importance to You	Rating of Vendor 1	Rating of Vendor 2
Speed of set up			
Quick alignment results			
Fast adjustment			
Tradesman reaction			
Durability			
Life cycle cost			
Understandable method			
Memory			
Training			
Other applications			
Total cost			
Periodic calibrations required			
Special Feature			
Vertical pumps			
Long spans			
Extra accessories			
Intrinsically safe			
Limited rotation			
Evaluation Scores			

Brochures are available for download. Communications is easy. The only problem with the site is that the descriptions of the product are too limited to make a decision to even download the brochure.

CONSULTANTS AND ENGINEERS

You would expect that many consultants have web sites for three reasons. The first reason is that the cost is low. The site can even be setup by a moderately computer savvy consultant on an off day. It is also a place that you can direct potential clients to get immediate information from case studies, articles, and downloadable proposals. The third reason is that a well-designed site (modest is okay) that has a presence in the search engines will attract some business. The ad is out there even when you're working on another job!

http://www.dhr.com DHR Maintenance engineering services of Brad Pitt and Jeff Caplan. They are technology consultants. Their excellent RCM work is what got them on this list. They provide train and consulting support in this area.

http://www.royjorgensen.com/home/index.html International maintenance management consultants and trainers for governments and large organizations. They specialize in facility management and transportation. The web site discusses the availability of training and extensive international work.

http://www.managementtechnologies.com Management Technology Inc. TPM consultants. This firm seems to be a custom programming house with a strong subspecialty in maintenance. Their success stories are just a list of clients (impressive list, but not much to the stories). With their ProMaint product, they might better be placed in the CMMS section.

http://www.projetech.com Projetech consultants install CMMS and are VAD for the PSDI MAXIMO product. Due to line loading or site resource needs, the site loaded very slowly. I almost gave up. They give the E-mail addresses for individuals in the company, nice touch. Includes small links section.

http://www.maintrainer.com Springfield Resources is the author's company. We are consultants and trainers to the maintenance function in all types of organizations. We conduct many types of projects including audits, TPM, work sampling, and cost reduction. Training is a major part of the business with description of seven different 2-day seminars available both live and as self-study courses. Site includes extensive jump links to maintenance management resources throughout the world. There is an archive of articles from the newsletter.

http://www.maxhon.com.hk/index1.htm This is a Pacific Rim mainte-nance management consultant specializing in RCM. Site includes search engine (for the site). As of the date of the visit (Fall 1998), much of the site is under construction. Site map and easy E-mail contact.

http://www.lce.com/index.html The web page of Life Cycle Engineering. The site is very quick due to few graphics. A very complete maintenance consultant site with complete descriptions of products and offerings. They have some unique offerings in turnkey maintenance documentation develop-ment, and CMMS projects.

http://www.sunbeltengineering.com/mis_lnk.htm This is a mechanical engineering company with involvement with maintenance projects. One interesting thing about the site is the ability to meet all of the engineers with photographs, resumes and personal E-mail addresses. The site has excellent industry links.

http://weber.u.washington.edu/~brennon/reichert/profile.html This is Paul Reichert's company site. He is an expert in CMMS installation. The site highlights his experience and has several articles about the installation process. This is a very useful type site because it is simple, to the point, quick, and has enough information to decide to pursue (or not).

http://www.techlabs.com/techlabs/ TechLabs is a general systems consul-tant that has a specialized practice in maintenance management. It is most comfortable in the large system, government, large industry fields. Has sig-nificant modeling and statistical analysis capability

http://www.thekinseygroup.com/bck-grnd.htm The Kinsey group is a new consultant that specializes in maintenance and reliability issues.

http://home.fuse.net/MMCS/ This is another personal consultant's web site. This is an excellent low-cost way to get some notice. James S. Cullen, 8770 Wellerstation Dr., Cincinnati, Ohio 45249

http://www.ultranet.ca/bmc/bmc.htm This is a Canadian consultant that specializes in maintenance management. Their products include:

Needs Analysis We help clients determine their needs and requirements for maintenance improvement, including CMMS and other aspects of the maintenance operation.

CMMS Selection We assist clients in writing RFP's, assist in software selection and evaluation, and contract negotiation.

Implementation We work with clients in setting up their implementation plan for maintenance improvement, often by focussing on CMMS start-up.

Project Management We work alongside clients in executing their plan for maintenance improvement, and we provide necessary expertise to monitor project progress and adjustment to changing conditions.

Work Force Effectiveness Working more efficiently is likely the first gain to be made in improving the maintenance department, but BMC helps clients increase the effectiveness of their work force through which their is opportunity for ongoing improvement.

Work Management Effectiveness How work activity is managed is critical to continuing improvement to the maintenance organization, ensuring that planning, scheduling, and administration of the work load gives the desired result.

Standards, Metrics, Coding, and Configuration Management We provide our clients with the basis to manage and control change within the maintenance organization, including changes to the equipment and facility base line, change in maintenance activity, and change in information support systems such as CMMS.

Inventory Management Inventory management is an important part of the maintenance operation. We help clients develop their inventory management strategy and policy so the maintenance department can be most effective.

GOVERNMENT INTERNET SITES

The government is the single biggest user of the Internet. Every branch, agency, and almost every bureau has a web site. The individual sites tend to be enormous (behind a very modest home page). The three branches, if you remember your social studies, are the executive (http://www.white-house.gov), judiciary (http://www.uscourts.gov) and the legislative (http://www.house.gov and http://senate.gov). One of the design specifications of the original Internet (remember ARPANet) was to provide a communications network for government business that would survive having large sections removed.

There were many grants to setup web sites for particular interests. Some of the most powerful resources for the maintenance community are DOD, NASA, Department of Commerce, Library of Congress and the various publishing arms. As with corporations, the sites reduce costs because fewer workers are required to answer the phone or send out brochures.

At their best, government sites offer quick access to government research, archives, information, people and the ability to conduct business. In short, the government uses of the Internet parallel the public use. Some of the most powerful uses (some are just being unveiled) are the ability to download forms, like IRS 1040, and the eventual ability to submit the same forms (now restricted). Other uses include being able to check your own social security information, and being able to join bidder's and contractor's lists. I could apply to be added to a Small Business Administration bidder's list to bid for government consulting jobs.

The Department of Defense (DOD) has hundreds of sites devoted to thousands of topics. Some of interest to the maintenance community are:

http://www.arpa.mil DARPA (Defense Advanced Research Projects Agency) is the granddaddy of the Internet. They were the original source of funds for the project 20 years ago. Interesting site if you are a fan of advanced defense projects and gossip.

http://www-usappc.hoffman.army.mil/ Army publications server. This site is for DOD use or for the use of any organization that can get an account. Civilians can go to NTIS, shown below. The Army site offered interesting statistics on the number of visits (5,520,790 in 1997). It also described the server (which is only a 133 MHz Pentium).

http://www.ntis.gov/databases/armypub.htm National Technical Information Service Army Manuals and Publications. The National Technical Information Service distributes Army technical manuals (TM's), field manuals (FM's), Army regulations (AR's), technical bulletins (TB), and other similar publications to the general public under arrangement with the U.S. Army Publishing Agency.

http://www.acq.osd.mil/ens/sh/ This DOD safety site has all of the statistics and information related to military safety. Great site for maintenance professionals interested in or responsible for safety.

One example were the aircraft safety home pages (which can be found at http://www.acq.osd.mil/ens/sh/faa.gif):

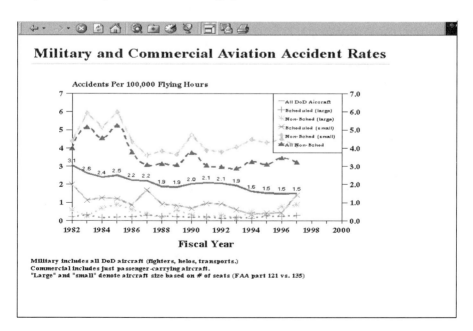

http://www.navy.fac.navy.mil/homepages/navfac Naval Facilities Engineering Command (the home of the Engineered Performance Stand-ards). This site is filled with good information for construction and engineering. They connect you to their acquisition system (you can find out what bids are available). Excellent source of links can be found at http://www.fac131.navfac.navy.mil/ies/hotlink. The NAVFAC maintains files of cost reduction ideas for use on bases throughout the world. The summer idea was:

☐ Save electricity in areas of adequate ambient light.
☐ Next time you have your vending machines serviced, have the technician remove the florescent lighting.
☐ Be sure to remove and make safe the ballast's as well as the lamps. A surprising amount of electricity can be saved on a typical base or building.

http://www.nrl.navy.mil/ The site for the Naval Research Laboratory. The NRL acts like a contractor research company for the Navy and other branches of the DOD. It has an entire section devoted to technology transfer.

http://www.acq.osd.mil/ec Information on electronic commerce, especially when in communication with the DOD. This site is the champion for E-commerce in the government. It has information, pilot projects, and advocacy activities. The idea is that E-commerce will help the military increase its efficiency.

NON-MILITARY GOVERNMENT SITES

http://www.doc.gov Department of Commerce has a deceptively simple home page. Like many other government sites, major resources lurk behind. The Department of Commerce is business's representative to the federal government. This site includes links to the census site, export sites, Bureau of statistics among others. A few of the popular links:

National Oceanic and Atmospheric Administration
> The NOAA Mission: To describe and predict changes in the Earth's environment, and conserve and manage wisely the Nation's coastal and marine resources to ensure sustainable economic opportunities.
> National Weather Service
> Nat'l Environmental Satellite, Data, and Information Service
> National Marine Fisheries Service
> National Ocean Service
> Office of Oceanic and Atmospheric Research

Coastal Ocean Program Office
Office of Global Programs
Nat'l Telecommunications & Information Administration

NTIA serves as the President's principal adviser on telecommunications policy and is responsible for domestic and international telecommunications and information technology issues, working to spur innovation, encourage competition, create jobs, and provide consumers with better quality telecommunications at lower prices.

Institute for Telecommunication Sciences is the research and engineering branch of NTIA.

Patent and Trademark Office

For over 200 years, the basic role of PTO has been to promote the progress of science and the useful arts by securing for limited times to authors and inventors the exclusive right to their writings and discoveries.

Technology Administration

TA manages 3 major agencies, the National Institute of Standards and Technology (NIST), the National Technical Information Service (NTIS), and the Office of Technology Policy (OTP).

National Technical Information Service

NTIS is the central resource for government-sponsored U.S. and worldwide scientific, technical, engineering, and business-related information. As a self-supporting agency of the Commerce Dept., NTIS covers its business and operating expenses with the sale of its products and services.

National Institute of Standards and Technology

NIST was established by Congress "to assist industry in the development of technology..." Its primary mission is to promote U.S. economic growth by working with industry to develop and apply technology, measurements, and standards.

http://www.doe.gov Department of Energy, energy grants are available from the Rebuild America program that provides energy efficiency grants for schools, commercial buildings, and hospitals.

http://www.eren.doe.gov. -Department of Energy information administration. These energy sites are hundreds of pages with numerous links to other resources. The range of topics that I could view spanned from weekly coal production reports by state and region to educational materials for elementary schools. A maintenance user might be interested in topics such as recycling, conservation, and new technologies.

http://www.dot.gov The Department of Transportation is responsible for everything that moves! Well almost, it doesn't cover little red wagons. For the maintenance professional, the DOT can be a great resource in information on mobile equipment. There are many research projects going on that might be of some use to you. To give you an idea about the scope of coverage, the home page lists:

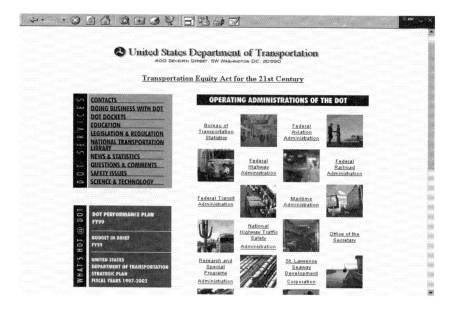

http://www.usda.gov The Department of Agriculture has one of the many powerhouse sites. It represents the interests of the farmers and the food business. It has 25 or more major bureaus. Those of you that are have small farms or ranches can get great free information from this site.

http://www.fda.gov The Food and Drug Administration bills itself as the foremost consumer protection association. From their home page, you can travel to many major branches including human drugs, cosmetics, and youth smoking, etc. Maintenance professionals wanting to learn more about medical topics, heath sciences, pharmaceuticals, veterinary science would find this site very interesting.

http://www.ed.gov Department of Education If you are interested in any area of education, this is a great starting place. The site includes research results, government actions, statistics, news, and statements from everybody who is anybody in government and education. Maintenance people in education could spend some time surfing the branches of this site.

http://www.hud.gov Department of Housing and Urban Development. The HUD site is devoted to the business of public housing. It includes resources for getting involved, information about programs, and the ability to download manuals and HUD guidelines. The site also has links to outside sites. Very useful for any maintenance professional with contact to housing, dorms, hotels or barracks.

http://www.usdoj.gov While not usually of direct interest to the typical maintenance professional, the Department of Justice site has significant resources. The FBI is an agency of the Department of Justice. You can spend a few minutes perusing their fugitive photos (surprise, the locksmith you just hired has opened one too many doors!) or visiting the ever interesting X-files (I looked but didn't find them). The other link that had interesting results was the Freedom of Information Act (FOIA). From there, you could download files on famous people investigated by the FBI, looked pretty interesting.

(http://www.dol.gov) Department of Labor is of interest to anyone in maintenance. The site featured stories about summer jobs, 401K plans, outside resources. They have an extensive on-line library, and there are links to about 20 agencies with the Department of Labor.

ADDITIONAL GOVERNMENT SITES OF INTEREST TO THE MAINTENANCE PROFESSIONAL

http://www.usdoj.gov/crt/ada/adahom1.htm Americans with Disabilities Act is a Justice Department site so the orientation is toward the legal aspects rather than the engineering or maintenance aspects of the ADA. It includes a complete resource guide to the ADA. Sections include the law, court cases, settlements, agencies, guidelines, and technical assistance.

http://www.gsa.gov The General Services Administration home page led to the PBS or Public Buildings Service. While the key word "maintenance" didn't hit anything in the GSA index, there was plenty of information on procurement, contracting, bidders, and construction. There was a link to the Virtual Library at http://www.arnet.gov/References/References.html This library had hundreds of lists, papers, and guidebooks about procurement.

http://www.epa.gov. One of the Environmental Protection Agency most popular programs has its own site, EPA Energy Star program http://epa.gov/docs/GCDOAR/energystar.html.

http://www.fema.gov/ Federal Emergency Management Agency FEMA is the agency responsible for emergency response for disasters throughout the U.S. It features many pages for small businesses to get some of the rebuilding work and other FEMA contracting opportunities. The links page has over 100 or more entries long for someone interested in looking deeper into the disaster preparation field. Each entry has a date that shows when it was last verified (an excellent idea).

http://www.usps.gov/ Postal Service It's pretty easy to get paranoid. When I entered the URL for this site the screen went blank for what seemed to be an excessive time. Then a huge graphic started to load (that took a long time, too). It turned out to be a 1.2 meg home page (for the novices out there that is about 20 times too much)! The site was useful for PO users wanting rates or to order supplies.

http://www.osha.gov/ The OSHA website is a very powerful. It allows you to look up OSHA rules or detailed inspection lists, and to check the inspection history for facilities throughout the U.S. by state, by company name. If you are responsible for safety then you should have this site bookmarked.

OSHA Office of Training and Education is in the Salt Lake City, OSHA center. The site had resources to learn about the laws, safety, reporting, centers for education and other useful things.

Office of Training & Education

OTI Bulletin - Reintroduction of Outreach Update Requirement
OTI Schedule of Courses
OTI Courses, Status & Availability (Training & Registration Database)
OTI Education Centers
Outreach Materials for Agency Initiatives & SEPs
OTI Bulletin - Reintroduction of Outreach Update Requirement
OTI Schedule of Courses
OTI Courses, Status & Availability (Training & Registration Database)
OTI Education Centers
Outreach Materials for Agency Initiatives & SEPs

http://www.sbaonline.sba.gov/ Small Business Administration online is the primary government resource for small businesses. The site features links to the SBA library, U.S. small business statistics, and the ability to sign up for bids. A great site if you want to start a business.

http://www.nist.gov The National Institute of Standards and Technology (NIST) is an agency of the Department of Commerce Technology Administration. The agency has primary accountability for discussion and communications of standards, quality, and manufacturing excellence. They maintain laboratories in most of the major sciences. National Institute of Standards also has a measurement site for calibration, certified reference programs, weights and measures and standard reference data products.

http://www.nist.gov/item/NIST_Measurement_Services.html. The NIST also has a manufacturing engineering laboratory at http://nist.gov/mel/mel-home.html.

SAMPLING OF MAJOR US GOVERNMENT INFORMATION SERVERS

http://www.edfacilities.org/index.html The National Clearinghouse on Educational Facilities (NCEF) is one of the Internet gems (if, of course, you manage schools or university buildings). This site has reams of information about the design, maintenance, and operation of school buildings. Their links page qualifies the site as a supersite: http://www.edfacilities.org/links.html. Other resources include:

Planning and Finance	Construction
Physical Design	ERIC Database
Internet Resources	Operations and Maintenance

http://www.loc.gov/ The Library of Congress is the ultimate library. There is information on every topic. You can spend centuries here looking for interesting things to read. Statistics for the site show that over a trillion bytes a day are transferred.

http://www.nsf.gov/home/pubinfo/start.htm The National Science Foundation is the federal foundation that funds science research. It has hundreds of resources for the browser.

http://hypatia.gsfc.nasa.gov/NASA_homepage.html NASA is one of the larger users of the web with hundreds of sites and millions of pages. You can find out about the space shuttle, planetary research, and find out where look at live videos of the current mission (it gives you details that you were waiting for - now at satellite GE2 transponder 9C, frequency is 3880 MHz with an orbital position of 85 degrees West Longitude, with audio at 6.8 MHz, in case you were interested).

http://www.fedworld.gov/ FedWorld Information Network This site is a gateway to 100 government bulletin boards, 20 government databases, many agencies, and statistics. It offers access to millions of pages of government sponsored or generated data. To find government jobs, surf to http://www.fedworld.gov/jobs/jobsearch.html

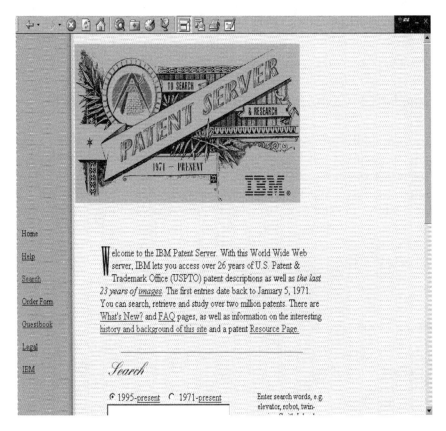

http://www.uspto.gov Patent and Trademark office. IBM also maintains another patent site server. The IBM site allows detailed searches back to 1970! **http://www.patents.ibm.com/ibm.htm/**

NEWSGROUPS, FORUMS, AND CHAT

What could be more interesting than having a chat with people all over the country (and world, for that matter) about a topic of vital interest to you? It is more and more common to have friends who spend hours a week reading and responding to postings on their favorite newsgroup. Fishermen, four wheelers, and model train enthusiasts will find active groups to join. On the business side, accountants, small businesspeople, and people who love (or hate) Microsoft will find several homes among the newsgroups.

NEWSGROUPS

Newsgroups are groups that are bound by a common love, hate, interest, or membership. By early 1998, there were over 30,000 Newsgroups on every conceivable topic. Topics range from people who collect stamps to people who hate particular politicians or fast food or love anagrams. There are several newsgroups devoted to engineering, buildings, and fleet, but few active ones on maintenance management.

The people who started the original newsgroups were devoted to their topic. These newsgroups were their lives. There were very strict limits to the commercial content, unrelated postings, and to anything that impeded the orderly exchange of ideas. To learn these rules, read the netiquette section of Chapter 2. Another recommendation is to follow the newsgroup for a few weeks to see what people post and how they respond. This watching is called net lurking.

To find newsgroups of interest, either use your newsreader (one comes with your browser) or lookup some of the groups mentioned in this chapter. Frequently, the first stop in the search for relevant newsgroups might be a familiar supersite. Finding useful sites for discussion in the maintenance world requires detective work.

The supersite called Vibrate.net features several newsgroups that are of interest to vibration enthusiasts (and are potential starting points for a maintenance newsgroup search). These include:

sci.engr

sci.engr.mech

sci.engr.analysis

sci.techniques.testing.misc

sci.techniques.testing.nondestructive

sci.engr.manufacturing

sci.engr.marine.hydrodynamics

sci.mech.fluids

Using the surfing ability of the supersite, you can jump to a discussion site of interest. Choose a likely candidate from the list on the supersite. a typical posting might look like this:

Here are another couple of little mysteries we may be able to clarify. I've often wondered about torques of big end bolts that are stressed to "elastic limit" when tightened. If they are indeed stressed to elastic limit, how do they survive the added loading (which is quite significant) of piston reversal of travel at TDC??(inertia reversal),one would tend to think the bolts would stretch beyond the "point of no return", loosen and, well, we know what follows this don't we? Also, I've noticed bolt torques(standard torques) are very similar between bolts of the same size and grade with differing thread pitches(coarse vs. fine). As a thread is effectively an inclined plane wrapped around a shaft, would this not imply a different "gearing" between threads of different pitches, resulting in higher tension value generated in fine thread bolt. Is the difference consumed in extra friction in the fine thread?

Pretty interesting stuff!

Another strategy is using the search engines. Many of them will search for phrases, keywords, authors, and companies in newsgroups. DejaNews (http://www.dejanews.com) is one of the better newsgroup search engines. A search in DejaNews under "maintenance management" turned up some interesting possibilities. Note the newsgroup is the second field from the right. Some of these are clearly of interest, others would have to be looked at more closely:

QUICK SEARCH RESULTS

Matches 1-25 of exactly 26 for search "maintenance management":

	Date	Scr	Subject	Newsgroup	Author
1.	98/06/15	047	Maint Management, Calibration	alt.business.misc	Robert Ruland
2.	98/06/15	047	Maint Management, Calibration	alt.business.seminars	MFG SOFTWARE
3.	98/06/15	047	Maint Management, Calibration	alt.comp.shareware	MFG SOFTWARE
4.	98/06/17	046	National Maintenance	gov.us.fed.congress.g	
5.	98/06/17	046	Educational Facility Conditi	misc.education	hutchisonj
6.	98/06/17	045	GAO Protest Valenzuela	gov.us.topic.law.pub-	Jerry Walz
7.	98/06/15	045	Re: Pentagon caught in lie	comp.software.year-20	T.S. Monk
8.	98/06/18	044	Re: Please don't reply to th	alt.comp.shareware.pr	Phill Hellewell
9.	98/06/15	044	Educational Facility Assessm	k12.ed.business	hutchisonj
10.	98/05/28	044	Found: Maintenance Managemen	alt.industrial	Joseph Hauber
11.	98/05/28	044	Found: Maintenance Managemen	ieee.general	Joseph Hauber
12.	98/05/28	044	Found: Maintenance Managemen	misc.industry.pulp-an	Joseph Hauber
13.	98/05/28	044	Found: Maintenance Managemen	misc.industry.utiliti	Joseph Hauber
14.	98/05/28	044	Found: Maintenance Managemen	sci.engr	Joseph Hauber
15.	98/05/28	044	Found: Maintenance Managemen	sci.engr.manufacturin	Joseph Hauber
16.	98/06/08	043	Maintenance Resource Page	sci.engr.mining	Alexander
17.	98/06/04	041	DalTech Continuing Education	ns.general	Parker Barringt
18.	98/06/01	041	Maint Management, Vendor Man	alt.business.misc	MFG SOFTWARE
19.	98/06/01	041	Maint Management, Calibration	alt.business.seminars	MFG SOFTWARE
20.	98/06/01	041	Maint Management, Calibration	alt.industrial	MFG SOFTWARE
21.	98/05/30	041	Maint Management, Calibration	alt.industrial	Robert Ruland
22.	98/05/30	041	Maint Management, Calibration	alt.manufacturing.mis	MFG SOFTWARE
23.	98/05/29	041	Maint Management, Calibration	alt.business.misc	Robert Ruland
24.	98/05/22	041	Maint Management, Calibration	alt.business.misc	Robert Ruland
25.	98/05/27	038	Re: So what's with Phills mem	alt.comp.shareware.pr	Mike Philben

The population of mail lists and newsgroups changes rapidly. It might take several tries to find a few that are concerned with the problems and issues that you face. If you find one, consider posting a request for other newsgroups or mail lists.

MAILING LISTS

A mailing list is a newsgroup that is conducted by E-mail. Any messages that are sent to a special E-mail address are then sent to the whole list (sometimes a moderator intervenes). There is a second address that handles the commands to manage the list (such as subscribe and unsubscribe)

There are two programs in common usage for mail lists. They are Listserv and Majordomo. To find out the Listserv command set, send an E-mail message to

mailto:LISTSERV@MSU.EDU

with no subject and add the command "INFO LISTSERVE" to body of message

To subscribe to a discussion group on manufacturing:

Send E-mail to mailto:LISTSERV@msu.edu

and leave the subject line blank.

In the body of the message put in the phrase "subscribe MFG-INFO (your name)"

Example of a mail list specializing in TPM

TPM mailing list subscription address:

mailto:SHULACK@EUROPA.COM

Newsgroups of possible interest to maintenance professionals

sci.engr.control Newsgroup for people interested in industrial controls

misc.industry.quality Newsgroup dedicated to discussions on the quality issue focusing on ISO 9000

sci.engr.heat-vent-ac An unmoderated group for discussion of the scientific aspects of HVAC

sci.engr.lighting Discussion group dedicated to discussions of issues related to all types of natural and manufactured light

sci.engr.manufacturing Discussion group focusing on the engineering science of manufacturing

sci.engr.safety Discussion group dedicated to all aspects of human safety.

Website based forum/discussion group

Reliability.com is a consulting organization in the area of RCM and provides training in reliability engineering. The forum they created is one of the more active ones.

RELIABILITY DISCUSSION GROUP

Welcome to our new discussion group where visitors can stop and ask a question on a maintenance or reliability topic or just browse the different messages and responses to see if it adds value to your workplace. Please let us know if there is any way that we can improve this site. It is here for you to share ideas with other people in the manufacturing community. Feel free to pass the word about this discussion group to your co-workers.

VIBRATION MONITORING EQUIPMENT - David Carpenter 09:04:56 6/30/98

BALL BEARINGS LIFE DISTRIBUTION - YANN BOUTIN 09:01:51 6/01/98 (2)

     ANSWER - Yann Boutin 21:15:23 6/06/98 (0)

      BALL BEARINGS LIFE DISTRIBUTION - Bill Marsh 12:29:56 6/02/98 (0)

RCM Implementation - Go it alone or with a consultant? - Rich 10:10:04 5/21/98 (2)

     RCM Implementation - Go it alone or with a consultant?

 - John Moubray 05:33:22 6/25/98 (0)

     RCM Implementation - Go it alone or with a consultant?

 - John Weigelt 06:47:04 6/20/98 (0)

Improve Reliability w/ Latest Advances in Industry

      - Sheila Kowalski @ Pacific Energy Association 13:31:32 5/06/98 (0)

Reliability Engineer Opening - Tim Brandt 14:06:12 5/01/98 (0)

Failure rate prediction - life study - Scott C. Bly 09:53:59 4/23/98 (1)

      Failure rate prediction -life study-Konstantin Kamberov 07:32:52 6/18/98 (0)

Failure data analysis. - Devesh Dahale 18:38:30 4/20/98 (2)

      Failure data analysis. - Bill Burns 10:51:20 4/22/98 (1)

      Failure data analysis. - C. Lucero 15:59:54 7/13/98 (0)

Root Cause Analysis in Healthcare - Jacquelyn Niesen 14:04:45 4/18/98 (0)

Root Cause Analysis in Healthcare - Jacquelyn Niesen 14:04:07 4/18/98 (1)

      Re: Root Cause Analysis in Healthcare - Patrice Spath 14:04:36 4/19/98 (0)

Interesting Reliability problem. - Devesh Dahale 17:31:50 4/10/98 (0)

Predictive maintenance info. req. - Devesh Dahale 17:25:45 4/10/98 (2)

      Predictive maintenance info. req. - Christian Stanciu 05:03:48 5/27/98 (1)

      Alignment Systems info. req. - Stanciu Cristian 05:09:10 5/27/98 (0)

Definition of: Limited Life Items - Christopher B. Lucero 10:09:36 3/18/98 (3)

      Definition of: Limited Life Items - Kirk Gray 11:11:42 3/20/98 (2)

Please...more opinions, though I agree with Kirk - C. Lucero 15:21:10 4/07/98 (1)

      Please...more opinions, though I agree with Kirk - Steven Needler
      16:56:09 4/15/98 (0)

RCM Software ? - Jon N. Didriksen 04:53:58 3/03/98 (0)

Know of any good Vibration Training? - Gary Fulmore 12:02:41 2/27/98 (3)

      Know of any good Vibration Training? - Wayne Tustin 14:48:42 4/14/98 (0)

Know of any good Vibration Training? - Kirk Gray 17:48:47 3/05/98 (0)

Know of any good Vibration Training? - Kirk Gray 17:37:46 3/05/98 (0)

Maintenance Management Terminology - Fred Sharpe 18:18:06 1/26/98 (1)

Maintenance Management Terminology - david rivera rubio 18:35:13 3/26/98

MRO storekeeper magazine - Julian Hart 07:49:16 9/24/97 (0)

Failure Rate, Prediction - for S/W - Shmuel Lederman 10:53:25 9/21/97 (1)

Failure Rate, Prediction - for S/W - Steven Needler 12:46:46 9/23/97 (0)

Electric Motor Predictive Maintenance - Charlie Martin 15:42:43 9/05/97 (8)

Electric Motor Predictive Maintenance-mike closson 21:14:48 10/20/97 (2)

Electric Motor Predictive Maintenance - Scott Bly 09:40:40 4/23/98 (1)

Electric Motor Predictive Maintenance- Robert Coffee 10:40:57 6/03/98 (0)

refinery maintenance case study - Garvie 13:59:01 9/01/97 (0)

Predictive/Preventive/Proactive Maintenance - Choon Ooi 12:06:02 8/26/97

Predictive/Preventive/Proactive Maintenance - david rivera rubio 18:40:33 3/26/98

Maintenance Planning Activities - HAMDI ABDULQADIR 07:54:56 8/23/97

CMMS Software - Chandra Locke 19:44:55 8/22/97 (0)

PdM Job Descriptions and Pay Scales - Randy Dallas 15:24:05 8/19/97

PdM Job Descriptions and Pay Scales - Jim Zapetis 12:13:27 5/19/98

RCM facilitators - Conrad Davies 04:47:44 7/29/97

RCM facilitators - John Moubray 05:46:17 6/25/98

RCM facilitators - Mike Espenschied 23:15:52 8/16/97

RCM facilitators - Jeff Caplan 14:31:47 8/08/97

RCM facilitators - Chris Shepley 18:20:09 1/25/98 (1)

RCM facilitators - Peter Ball 06:13:55 4/13/98 (0)

spare parts - Alex Green 07:34:30 7/28/97 (0)

spare parts - Bndicte Retaileau 15:14:28 7/23/97 (5)

spare parts - Alex Green 02:24:53 7/29/97 (0)

reliability risk - corey crowley 11:06:27 7/21/97 (1)

reliability risk - David Coit 11:53:22 7/23/97 (0)

EQUIPMENT AND SYSTEM RELIABILITY AND AVAILABILITY

Marcos Monteferrante Farina 15:24:17 7/04/97 (2)

EQUIPMENT AND SYSTEM RELIABILITY AND AVAILABILITY - John Doe 16:08:33 8/22/97 (0)

EQUIPMENT AND SYSTEM RELIABILITY AND AVAILABILITY - Graham Parker 06:40:10 7/24/97 (0)

For information on Reliability Center, Inc.'s products and services, please send e-mail to "mailto:info@reliability.com" info@reliability.com, phone us at (804)458-0645, or FAX request to (804)452-2119.

Not to be confused with the previous site, another reliability forum can be found at Reliability Magazine. Reliability magazine has a moderately active discussion group in the area of maintenance management. Some of the threads of interest can be found at the following:

http://www.reliability-magazine.com/wwwboard/wwwboard_gen/messages/60.html

http://www.reliabilitymagazine.com/wwwboard/wwwboard_rel//wwwboard.html

User groups make the CMMS or PdM package you buy more valuable. Do you own a CMMS and want to talk to others with the same system? Many user groups are going on-line as newsgroups. Here you can read comments about the software, ask questions of the whole group, get help, and gripe to your heart's content. Some user groups are private (those sponsored by CMMS sites), accessible only through the host vendor's site and require a password.

http://www.dingos.com/forum.htm

A new forum in the area of oil analysis. One of the larger newsgroups/forums is attached to the Entek/IRD site.

ENTEK IRD CUSTOMER SUPPORT BULLETIN BOARD

This Bulletin Board was designed to be a discussion forum for users in the Vibrations/Predictive Maintenance field to freely ask and answer questions. Since it is not guaranteed to be monitored by an Entek IRD Customer Support Specialist at all times, if you have a situation with any of Entek IRD's Software or Hardware, please contact us via email at our to add a Bulletin Board message.

---

http://www.uesystems.com/uepost The UE forum has postings for people interested in predictive maintenance. There is a button that links to information that you can read about how to post messages. Note: you may need to reload this page to see the most recent additions.

---

air lines and compressors andy bogan my e-mail address is VWbus74.aol.com 06 Feb 1998
> air lines and compressors C. DeBaer 08 Feb 1998
> air lines and compressors Mike Pavuk 02 Mar 1998
> air lines and compressors "The High Cost of Procrastination"
> Dana J. Payton 01 May 1998
> air lines and compressors "The High Cost of Procrastina...

Terrence O'Hanlon 11 May 1998

air lines and compressors Terrence O'Hanlon 11 May 1998

ULTRAPROBE 100 Barry Silken 16 Feb 1998

Bearing defect detection using UE2000 probe Yvan A. Lejeune 16 Feb 1998

Bearing defect detection using UE2000 probe Yvan A. Lejeune 16 Feb 1998

Instrument for measuring the flow of gas and liquids Domen Camlek 18 Feb 1998

Instrument for measuring the flow of gas and liquids Mike Pavuk 02 Mar 98

Software Applications Jeff Lottes 20 Feb 1998

Software Applications Cyndie Hinds 28 Feb 1998

Software Applications Ivan Cyr 03 Mar 1998

Software Applications Cyndie Hinds 14 Mar 1998

Software Applications Liane Harris 31 Mar 1998

---

http://ll.nfesc.navy.mil/leletoc.htm One of the most interesting forums is the U.S. Navy (NAVFAC -Naval Facilities Engineering Command) lessons learned forum. In this forum Navy personnel discuss lessons they learned doing various projects. In one section they located and printed links to other lesson's learned pages:

Australian Environmental Resource Information Network

Agency for Toxic Substances and Disease Registry (ASTDR)

Consumer Product Safety Commission (CPSC)

Department Standards Committee "http://www.dsc.doe.gov/llearned.html"

Environmental Protection Agency "http://www.epa.gov/"

FinanceNet "http://www.financenet.gov/"

International Atomic Energy Agency (IAEA) World Atom http://www.iaea.or.at/worldatom
NASA Lessons Learned Information System (LLIS)

"http://envnet.gsfc.nasa.gov/ll/llhomepage.html"

NASA Selected Current Aerospace Notices (SCAN)

"http://www.sti.nasa.gov/scan/scan.html"

OSHA Standard Interpretations

"http://www.osha-slc.gov/OshDoc/Interp_toc/Interp_toc_by_std.html"

U.S. Geological Survey (USGS) Natural Hazards Programs: Lessons Learned for
Reducing Risk   "http://h2o.usgs.gov/public/wid/html/HRDS.html"

AirForceLINK Factsheets (Air Force Lessons Learned)
"http://www.af.mil/news/indexpages/fs_index.html"

Automated Lessons Learned Collection and Retrieval System (ALLCARS)
"http://www.afam.wpafb.af.mil/prod03.htm"

Center for Army Lessons Learned (CALL) "http://call.army.mil:1100/call.html"

Department of Defense Technical Information Web "http://www.dtic.dla.mil/dtiw/"

Department of Defense Technology Transfer (TechTransit)
"http://www.dtic.dla.mil/techtransit/"

Federal Technology Transfer
"http://www.dtic.dla.mil/techtransit/techtransfer/fed_t_2.html"
Naval Facilities Engineering Command (NAVFA C) Lessons Learned Home Page
"http://ll.nfesc.navy.mil/"
Product Acquisition and Engineering (Navy Lessons Learned)
"http://prodeng.nwscc.sea06.navy.mil/default.htm"
Commercial Educational "http://www.ul.com/auth/tca/v5n2/index.html"
The Code Authority - Underwriters Laboratories Newsletter
Internet Disaster Information Network "http://www.disaster.net"
EnviroWebs Internet Listing of Environmental Information Services
"http://www.envirolink.org/EnviroLink_Library/"
Electric Power Research Institute (EPRI) "http://www.epri.com/"
Federal Emergancy Management (FEMA)s Global Emergency Management System
"http://www.fema.gov/fema/gems.html"
Mayo Clinic and IVI Publishing Inc .Online Health network "http://healthnet.ivi.com"
Maintenance Management "http://www.efn.org/~franka/MMeasier/TIPS.html"
Medical Links (General) "http://www.netins.net/mega/me_menu.html"
Medical/Health Locations (via World Health Organization)
"http://www.who.ch/others/OtherHealthWeb.html"
Quality Management "http://www.qualinet.com/isopage.htm"
Risk Management and Insurance (RISKWeb) "http://www.riskweb.com"
Total Quality Management Resource Center Lobby (David Butler Associates, Inc.)
"http://www.zoom.com/dba/"
Lessons Learned Implementing a Navigation Server for the Web
"http://www.ncsa.uiuc.edu/SDG/IT94/Proceedings/HCI/glazer/glazer.html"
Quality Distance Education (QDE) "http://www.uwex.edu/disted/qde/home.html"

Newsgroups, discussion groups, mail lists are powerful ways to participate in a field. They provide an important bond between the new people entering the field and the old timers. The groups also provide a community for people with common interests. As you can see from this chapter the maintenance profession is in its infancy in the use of these types of resources. Expect a small explosion of forums, newsgroups and mail lists in the next year or two as more and more maintenance professionals get wired.

CONDUCTING RESEARCH ON THE INTERNET

Research and exchange of information are two of the original missions of the Internet. The Internet is designed to aid research. Before we get into the research though, it would be irresponsible not to repeat the Internet research rule: get independent verification of all Internet facts. In other words, don't bet your life savings on a tip you saw while researching mutual funds! There are terabytes of good information out there. This chapter will help you start the search.

All research starts with a question. The Internet is uniquely suitable to provide resources to answer certain types of questions. Maintenance issues are not a major focus for the Internet as are the fields of software or engineering yet even without the focus, a basic search on "maintenance management" identifies 200,000 resources!

TYPES OF RESEARCH ON THE INTERNET

Internet research almost always starts with a query to a search engine or a visit to a supersite. In the case of the search engine, knowing where to start is often half the battle. The second half of the battle is knowing how to structure the query to get the fewest overall results of the highest relevance.

The Internet is the best place for certain types of research. Some of the areas that the Internet shines are:

1. Locating any kind of product. Every day more manufacturers are putting their catalogs on-line. You can get answers to questions almost immediately. Most sites have easy E-mail hook-ups that allow communication with someone on the inside. Companies with serious commitment to the on-line world respond to E-mail within a day.

2. Background research is very quick on the Internet. When an organization wants to start a TPM or RCM program, the Internet can provide background material to start the discussion. Available materials include articles, products used in that field, software, consultants, user stories, and university research on the topic.

3. Technical problems can be looked up in manufacturers' databases or, if the problem is more generic, in university databases.

4. Issues relating to being a novice with a technology, software packages or machine. The primary research resources are the FAQ (frequently asked question) files. These ask and answer the novice questions.

5. Esoteric areas, advanced technologies and exotic materials have long been at home on the Internet. The Internet was designed to allow government and university laboratories to exchange this type of information. In short, the Internet is the home to the exotic and the obscure.

6. Any software issue at all. The software industry was the first commercial industry to adopt the Internet and the BBS before that for support, upgrades, demos and new products. You can find plug-ins, discussion groups, and user forums for most major products. If you need a product quickly, it is probably available as a download (using your trusty credit card).

7. Finding books and CDs. Great resources are available on the Internet if your research requires a particular book or CD.

Very early in my experience with the Internet I was looking for some support material on the topic of maintenance planning. A quick search of the net located an article, published at one of the government sites, that began:

"Maintenance Planning is the process used to develop anticipated maintenance requirements for a system and propose who will perform required maintenance tasks and where they will be performed...The goal of maintenance planning is to establish a management and planning process for achieving an equipment maintenance/repair capability to:

- ☐ Provide a logistically supportable and supported system as it deploys
- ☐ Ensure responsive support to using command operational requirements
- ☐ Reduce the risk of incurring unnecessary operating and support costs
- ☐ Ensure the timely availability of the required maintenance/repair support capability with the system it supports"

The document went on and had a complete bibliography. I had struck gold. Now I was hooked on Internet research.

Try this: Once on the Internet, research a topic of interest to you (whether related to maintenance or not).

EXAMPLE: RESEARCH ON TRIBOLOGY

Procedure: Enter "tribology" on Excite's search screen. The search locates over 2000 web sites where the word tribology is used (the first sites displayed usually have frequent use of the search word). The search engine lends me a hand by displaying a list of related words such as lubrication, ASME, lubricant, friction, hydrodynamic, misalignment, lubricant, greases, and fretting.

Since we are doing research I use my mouse to choose one that says:

67% WWW Tribology Info - Academic Links - Division of Tribology, Fachhochschule, Hamburg. Home of the Tribology Discussion List, maintained by Erik Kuhn. Institute for Production Technologies and Metal Forming Machine Tools, Technical University at Darmstadt. http://www.shef.ac.uk/~mpe/tribology/links/aclink.html

---

This site has academic links all over the world. Just the USA links fill a page! They include:

Advanced Coating and Surface Engineering Laboratory (ACSEL), Colorado School of Mines.
Center for Advanced Tribology, Western Michigan University.
Topographic Research and Analysis Lab. (TRAL), Worcester Polytechnic Institute
Wright-Patterson Air Force Base
Analytical Chemistry, Lubrication, and Tribology
Tribology at Argonne National Laboratory
Tribology (Friction, Lubricants, and Wear of Surfaces) and
    Energy Technology Division
Tribology Section, Naval Research Lab
Tribology Research, Georgia Tech.
Plasma Research Engineering Laboratory, Colorado State, USA
    Ion beam and Tribology Research maintained by John Davis.
Iowa State University Mechanical Engineering Department. Tribology Laboratory Details.
Tribology Research, Pennsylvania State University.
Purdue University.
    Mechanical Engineering Tribology Laboratory (Sadeghi).
Institute for Information Storage Technology, Santa Clara University.
Center for Advanced Friction Studies, Southern Illinois University, Carbondale.

MIT, USA.
    Tribology research (Suh)
The National Steel Technical Center, USA.

For extras they included a list of tribology related (places where tribology is studied or where research is going on) sites.

University of California, Berkeley, USA.
    Tribology in the Computer Mechanics Lab (Bogy and Komvopolous).
Talke Group Home Page, at University of California, San Diego.
    Tribology of disk and tape systems (Talke and Lauer).
3D Surface Topography
    An excellent set of reference pages maintained by Sullivan at William
    States Lee College of Engineering, NC, USA.
Tribology Research, Texas Tech University.
Tribology Laboratory, University of Akron
Bogey's Tribology Group, University of California, Berkeley.
Center for Magnetic Recording Research, University of California, San Diego.
Talke Group Home Page
Adair Research Group, University of Florida.
Tribology Lab, University of Illinois, Urbana-Champaign
Tribology lab, University of Michigan.
Surface Science Lab, University of New York, Binghamton.
Tribology/Manufacturing Laboratory, University of Notre Dame.
Tribology Lab, University of Pittsburgh.
Vibrations and Tribology Lab, University of Southern Florida.
Worcester Polytechnic Institute Topographic Research and Analysis Lab.(TRAL).
    Details of surface topography research and analysis at Worcester
    Polytechnic Institute, USA (Brown).

That little list should keep you going for a year or more. There are only 1999 web sites to go from the original search.

STRATEGIES FOR RESEARCH

The Internet lends itself to research because its roots were in universities. It was designed for researchers to share documents, files, and ideas across the world. The techniques for doing research vary depending on the obscurity of the references sought (in some ways the more obscure, the better). If you want to engage in serious research for a thesis or important presentation, then a quick visit to a library or bookstore would pay off. There are excellent books on Internet research. Be sure to look at the date of publication: the more recent is often better. You can also search the Internet itself for information on conducting research.

Let me repeat the caveat mentioned at the beginning of the chapter. *Independently verify any results you find.* Garbage, forgery, spoofing, unsubstantiated opinion, and just plain inaccuracy are commonplace on the Internet.

Designing the query is the most important part of initial research. The second most important issue is choosing the search engine and choosing the most likely place for the information. Specific rules for each search engine are covered in Chapter 4. Results from some informal surveys are also in that chapter (number of hits on a maintenance topic and relevancy of the first 25 sites).

If you were researching the opinions of people about the Microsoft Office Suite, then a search of the Usenet (newsgroups) would result in a mother lode of gold for sifting. A search for maintenance system vendors would be best conducted on the WWW where they are likely to have web pages. If you seek a clarification about rules relating to some esoteric sub-specialty of vibration analysis, then the traditional search of Gopherspace would be in order.

Start your search with as specific a query as will get results. Widen the search when you start to run out of sites or references. When searching the web for very specific information, use a spider site. If you seek a company or major concept, use the directory sites. In both cases you'll find super-sites that will help you locate additional resources. In general, engineering topics are better covered than maintenance topics.

ENGINEERING DESIGN RESOURCES

http://www.ecs.umass.edu/mie/labs/mda/dlib/dlib.html A University of Massachusetts library for support of undergraduate design projects. Links to their engineering department. Excellent information on engineering standards.

http://industrysearch.com IndustrySearch is a database of more than 350,000 manufacturing companies and 620,000 industrial suppliers. The site is designed for product design, plant maintenance, engineering and purchasing personnel.

http://www.dig.bris.ac.uk/hbook/ The Multi-Media Handbook for Engineering Design. The home page explains it all:

Multi-Media Handbook for Engineering Design
Design Information Group, University of Bristol

The Multi-Media Handbook for Engineering Design is a hyper-media database providing information relating to machine design. It provides university students with a concise source of key information giving the user quick and easy access to elementary engineering design principles, design details of machine elements and specific component information. It provides:

- design guides for a variety of design situations including the design, selection and application of components and systems

- catalogue information from component manufacturers to provide standard sizes and dimensions, ratings and capacities

- good practice guides to the proper design of components and systems in terms of increased strength, reduced cost, more efficient manufacture and assembly

- materials data for common engineering materials including properties, standard forms of supply, special treatments and typical applications

The provision of the handbook will aid students in their search for information and will provide support for the design lecturer in their teaching.

The handbook is authored in Windows based software called RV2 which is also product of the Design Information Group. The system combines database, hypertext and graphics features in a unique hybrid which was originally conceived as a means of improving access to engineering design data and has proved to be excellent for developing design guides. The flexibility of RV2 allows information to be added to the database at any time and for it to be automatically integrated thus enabling the database to be tailored to specific engineering design courses by the course tutor. RV2 includes its own editor to make authoring a simple matter. At present, 11 guides are available. These are:

- Casting
- Data Book
- Fasteners
- Roller Chain Drives
- Gears
- Springs
- Steels
- Rolling Bearings
- Transmission Selection
- Plastics & Rubbers
- Inverted Tooth Chain Drives

The handbook is available free of charge to all UK Higher Education Institutions. At present copyright restrictions do not allow us to distribute the handbook worldwide. Further details can be obtains by E-Mailing julian.cooke@bristol.ac.uk. As an example of the handbook this guide to Rolling Element Bearings is a WWW implementation of the equivalent handbook guide.

Acknowledgements

The Handbook has been developed during a two year project funded by the New Technologies Initiative (NTI) of the Joint Information Systems Committee (JISC is part of the HEFC).

http://www.ndltd.org The National Digital Library of Theses and Dissertations (NDLTD) was designed to improve graduate education by sharing the information by graduate students throughout the U.S.

http://www-jmd.engr.ucdavis.edu/jmd/ The American Society of Mechanical Engineers (ASME) publishes the *Journal of Mechanical Design* (JMD). This server provides information on the scope and purpose of the journal, the editorial staff, and information for authors. There is also an index to previous volumes and articles. Forms for submitting and reviewing technical papers are provided. Links to UC Davis Advanced Highway Maintenance and Construction Technology Center, a site dedicted to automation of highway maintenance (**http://www-ahmct.engr.ucdavis.edu/ahmct/**).

http://www.designinfo.com Very complete set of catalogs for mechanical design, with parametric search. Their site lets you:

- ☐ Search hundreds of on-line supplier catalogs simultaneously
- ☐ Compare hundreds of thousands of products
- ☐ Access product data in seconds

http://saviac.usae.bah.com This server is a central information resource for government activities, contractors, and academics concerned with structural dynamic analysis, design and testing, and related physical environmental effects. One could say the site is shocking (the name is Shock and Vibration Analysis Center)

http://www.civeng.carleton.ca/Other_Info/Other_Information.html This server located at Carleton University in Canada provides accounts of real engineering work written for use in engineering education. There are cases from almost all disciplines of engineering. The full text version of the catalog is also available via anonymous ftp as ASCII text or as a WordPerfect file from: **alfred.carleton.ca in directory /pub/civeng/ECL/**

http://www.memagazine.org/index.html   Mechanical   Engineering Magazine is sponsored by the ASME. Back issues available. Searchable index. Has a forum for members.

http://mecheng.asme.org   This ASME-sponsored database is a fully indexed, searchable system, with descriptions of more than 6000 engineering programs. As a web-based system, the database supports any number of direct URL references to your home site, email address, or other links you choose. You can add files to the public FTP site such as demonstration versions of your product, shareware files, documentation, or other files you would like associated with your software record. All files submitted to the archive, as well as the software database records, are published on CDROM for even wider distribution.

http://www.engr.utk.edu/mrc/ Maintenance and Reliability Center at the University of Tennessee, Knoxville. This site is dedicated to information sharing about maintenance management. You can see from their statement on the site:

Vision

The MRC will become the international focus for education, research, development, information, and application of advanced maintenance and reliability engineering.

A New Resource

The Maintenance and Reliability Center is a new and vitally important resource for industry. Headquartered at The University of Tennessee, the Center uses research and cutting-edge technology to help its member companies reduce losses caused by equipment downtime.

An Urgent Problem

Once perceived as a "practitioner" or manufacturing issue, maintenance engineering is now considered a business issue of urgent priority. Studies indicate that American industry spends more than $200 billion each year on maintenance. Production losses due to equipment downtime usually equal or exceed maintenance costs. Sixty percent of maintenance is unplanned.

Mission

The MRC is a direct response to these problems. The Center's fourfold mission emphasizes its commitment to improving productivity and profitability for its partners in industry.

Education

The University of Tennessee offers students certification in maintenance and reliability engineering. Students achieve this specialization while earning a degree in one of the standard engineering disciplines. Certification in Maintenance and Reliability Engineering can be earned with engineering degrees ranging from BS to MS and Ph.D. Summer internships are available on a competitive basis. The MRC has recently been awarded an NSF Funded Combined Research Curriculum Development program.

Research And Technology Development

The MRC conducts research and development on specific projects recommended by its industrial members. Projects are conducted by UT students and faculty at The University of Tennessee, and with collaborating universities such as The University of Alabama and national laboratories such as Oak Ridge National Laboratory and Pacific Northwest National Laboratory.

Information

The MRC sponsors regular technology conferences and workshops and publishes a quarterly newsletter highlighting current developments in the field. MARCON97 and MARCON98 were held in May of 1997and 1998 with great success. The next Maintenance and Reliability Conference will be held the week of May 10, 1999 in Knoxville, Tennessee. An announcement and "Call for Papers" will be posted in the near future.

Business Applications

The MRC develops benchmarking standards and modeling techniques to help companies evaluate their maintenance and reliability operations. The University of Tennessee's College of Business Administration provides a vital link to the business community through graduate education and executive training programs.

---

www.glue.umd.edu/enre/reinfo.htm National Information Center for Reliability Engineering, University of Maryland. Significant resources in the reliability area. On the page for Tools they list (**http://www.enre.umd.edu/ rmp.htm**) over 25 non-CMMS analysis programs.

ELECTRIC MOTORS MAKE THE WORLD GO AROUND. WANT TO FIND OUT MORE?

http://www.Pmdi.com/php2.php Precision MicroDynamics provides a CGI program/resource to help size DC gear motor systems. Users fill out a form that details the load characteristics and motor characteristics among other things. By pressing the "submit" button, the form processor takes over and performs the computations. A second level of analysis provides the torque/speed, torque/power, torque/efficiency and torque/current plots. These can be downloaded to your computer simply by saving the GIF that is generated.

COMPLETE LIBRARIES THAT CAN ANSWER YOUR MAINTENANCE RESEARCH QUESTIONS

http://www.ichange.com An AT&T service that has 1000 industrial trade journals and newspapers and links to the Thomas Register of North American manufacturers.

http://www.eiq.com A new web based information and research resource. It accesses 12,000 journals and a table of contents with over 7,000,000 articles in the technical, business, medical and scientific areas. This is a fee service and offers a 30-day trial.

http://solstice.crest.org/online/virtual-library/Vlib-energy.html The WWW virtual library on energy. This is an example of a grant-sponsored site.

Welcome to GEM, the Global Energy Marketplace

This powerful, on-line, searchable database of more than 2500 energy efficiency and renewable energy annotated Web links has been sponsored by the US Environmental Protection Agency and created by CREST to promote a more sustainable energy future and mitigate global climate change that results from energy use. You will find highly useful case studies, reports, publications, economic analyses, product directories, discussion groups, country profiles, mitigation assessments, and other beneficial resources.

---

http://www.nist.gov/welcome.html National Institute of Standards and Technology provides you with access to a broad variety of scientific and technical information resources in different electronic formats, from online information to CD-ROMs to databases and electronic documents.

http://www.nist.gov/srd/srd.htm National Institute of Standards and Technology Standard Reference Database. Contains over 50 databases in many technical and manufacturing areas.

http://techreports.larc.nasa.gov/cgi-bin/ntrs This server is an experimental service that allows simultaneous searching of various NASA abstract and technical report services. The service accesses many large research databases:

Ames Research Center	Langley Research Center
Astronomy & Astrophysics (ADS)	LANL Astrophysics e-Prints (ADS)
Dryden Flight Research Center	Lewis Research Center
Goddard Institute for Space Studies	Marshall Space Flight Center
Goddard Space Flight Center	NACA Reports (abstracts only)
ICASE	NACA Reports (full text)
Jet Propulsion Laboratory	Physics and Geophysics (ADS)
Johnson Space Center	Space Instrumentation (ADS)
Kennedy Space Center	Stennis Space Center

http://www.w3.org/hypertext/DataSources/bySubject/Overview.html
WWW virtual library includes an Engineering section.

http://ctca.unb.ca/CTCA/sources/ Contains sites of value to those in the AEC (Architecture, Engineering, and Construction) industry in Canada.

http://www.cybertown.com/campeng.html A virtual campus in a virtual town with links to engineering resources throughout the US. The campus is part of a virtual city.

MACHINERY INFORMATION SITES

http://www.automationnet.com Database of vendors, consultants, and systems integrators in the field of engineering automation under the subject Mechanical Engineering on-line Services. They say: "The AutomationNET is a central information forum for anyone who makes automation products, integrates automation systems, needs automated systems or for anyone who just wants to learn more about the automation industry."

http://www.hoppenstedt.de/componet The PT Primer provides information in the Adobe Acrobat PDF format that was reprinted from the Power Transmission Handbook (The Power Transmission Distributors Association, 1993). It contains two sections excerpted from the original 14 chapters.

http://www.iglou.com/pitt A webzine on process pumps and filtration with articles, on-line principle and fundamental catalogs, new product and technology introductions, and engineering educational and resource materials.

http://part.net This server offers information about a prototype service for finding parts via the WWW. This is a fascinating idea. The PartNET company builds a virtual web site where all of your parts suppliers seem to be. One click entry to their systems will reduce the costs of part acquisition. They say about their service:

Who is *Part*NET?

*Part*NET is a software development company that specializes in Distributed Internet Commerce™ for large buyers and sellers. *Part*NET eCommerce™ catalogs allow multiple suppliers to be presented through a single web address.

PM (tribology -the study of lubrication)

http://www.shef.ac.uk/~mpe/mattrib/tribology/ Almost every item of machinery has moving parts, bearings, gears, slides, seals and many others. The successful operation of moving mechanical equipment is dependent on the smooth running and long life of these parts. Research in the field of tribology has paved the way for reliability and longevity in industrial machines. Lists: Conferences, Jobs, Journals, ImEechE Design data guides, etc.

Reliability (in some ways we are all in the reliability business)

http://rome.iitri.com/rac/ The Reliability Analysis Center is an Information Analysis Center sponsored by DTIC, the Defense Technical Information Center. RAC's charter is to collect, analyze, and disseminate data and information to improve the reliability and maintainability of components and systems. IIT Research Institute has operated RAC since its inception in 1968.

MISCELLANEOUS SITES OF INTEREST

We've had a great trip through the Internet. This last chapter could easily be millions of pages long. Every visit generates more sites of interest. You know you been surfing too long if your favorites list scrolls off the screen! The Internet bug has taken hold when you start searching for things you need for work, and you run into things that you want to look into and you know if you follow too many of them, hours will go by, without the original work being done. Have a great time surfing.

JOB SEARCHING

One of the best uses of the Internet is job searching. There are thousands of jobs advertised on the net. A simple Infoseek search on "jobs" yielded 5,900,000 hits! A more detailed search on jobs + "maintenance management" yielded 19,000 hits. Some of them looked like this:

MP2 Maintenance Management Software Installation, Implementation, & Training Workforce Development Housing Maintenance & Management. U.S. Army Family Housing Units. We operated and maintained 150 housing units ...72% http://www.alaska.net/~pmc/services/ (Size 6.7K) Document Date: 1 May 1998

Personalize Help - Check Email Home: Business and Economy: Companies: Computers: Software: Business: Business Management:..69% http://www.yahoo.com/Business_and_Economy/Companies/Computers/ Software/Business/Business_Management/Maintenance/ (Size 9.1K) Document Date: 30 Jun 1998

[Follow Ups] [Post Follow-up] [ATMS Online Forum] Posted by Mohamed

Taher on August 16, 1997 at 06:29:50: In Reply to: Maintenance Management Systems posted by Brian Olsen on July 15, 65% Document Date: 16 Aug 1997

http://www.itsonline.com/atmsforum/messages/143.html (Size 3.5K) 210-662-8264 (H). Current Employment. Presently employed by West Telemarketing Outbound, San Antonio, Texas, as an Applications Analyst. Responsible for developing and 61% Document Date: 2 Feb 1996 http://www.lookup.com/homepages/57428/resume.html (Size 4.5K)

DEDHAM, MASS. (July 21) BUSINESS WIRE -July 21, 1998--The leading edge manufacturing companies implementing enterprise asset management (EAM) software ...60% http://www.pathfinder.com/money/latest/press/BU/1998Jul21/891.html (Size 7.9K) Document Date: 21 Jul 1998

DISTRIBUTION AND PROPERTY MANAGEMENT BRANCH. MAINTENANCE MANAGEMENT BRANCH. Functions. a. Develop and manage the BILI supply program; develop and implement appropriate ADP programs relating to 59% Date: 27 Jul 1998 http://www.monmouth.army.mil/cecom/lrc/leo/plmdiv/mmbr.html (Size 5.9K) Document

---

Almost every CMMS, PdM, and maintenance company is advertising for different types of job openings. If you are looking or are thinking of looking for a job, the Internet is like having a thousand classified ad sections. One place to start your search is at the site of the CMMS, PdM, distributors, or tool vendors, that you now use and enjoy.

FOR WEB DESIGNERS

There is research being carried out about what works and what doesn't work on the World Wide Web. Scientists have observed people surfing the web and note what they look at, how long they look and what they retain. One of the most outspoken experts on this topic is Jakob Nielsen. His site http://www.useit.com is a must visit for web designers. In an interview by Katharine Mieszkowski, for an article *Usability Makes The Web Click*, which first appeared in the webzine, Fast Company (http://www.fastcompany.com issue 18) page 56, Nielsen offered some easy-to-implement ideas for improving Website design.

What's wrong with Web design?

Too many Web designers substitute a marketing agenda for a focus on what customers want. Users want speed, utility, and credibility - not portals, banners, or even community. And speed is the overriding criterion: Minimalist design rules. One phrase sums up the dominant mentality of the Web user: "I'm driving." People don't spend lots of time on any one page, because in order to feel that they're accomplishing something, they have to keep moving. The best kind of site shows users what each page is about and then quickly gets them to the next page.

Why don't more sites work that way?

Most developers fail to treat the Web as a new medium with new rules. The dominant metaphor is TV - think "channel," "show," and "eyeballs." But the Web is an interactive, one-to-one medium in which everyone can be a producer or a publisher. It isn't like newspapers or magazines either. At IBM and at Sun, we studied how people read on the Web. What we discovered is - they don't! They scan. Only 16% of Web users actually read word by word. So, on any given topic, people should write about half as many words for the Web as they would for the printed page.

Top Ten Mistakes in Web Design

1. Using Frames

Splitting a page into frames is very confusing for users. All of a sudden, you cannot bookmark the current page and return to it (the bookmark points to another version of the frameset), URLs stop working, and printouts become difficult.

2. Gratuitous Use of Bleeding-Edge Technology

Don't try to attract users to your site by bragging about use of the latest web technology. You may attract a few nerds, but mainstream users will care more about useful content and your ability to offer good customer service. Unless you are in the business of selling Internet products or services, it is better to wait until some experience has been gained with respect to the appropriate ways of using new techniques. When desktop publishing was young, people put twenty fonts in their documents: let's avoid similar design bloat on the Web.

3. Scrolling Text, Marquees, and Constantly Running Animations

Never include page elements that move incessantly. Moving images have an overpowering effect on the human peripheral vision. A web page should not emulate Times Square in New York City in its constant attack on the human senses: give your user some peace and quiet to actually read the text!

4. Complex URLs

Even though machine-level addressing like the URL should never have been exposed in the user interface, it is there and we have found that users actually try to decode the URLs of pages to infer the structure of web sites. Users do this because of the horrifying lack of support for navigation and sense of location in current web browsers. Thus, a URL should contain human-readable directory and file names that reflect the nature of the information space.

Also, users sometimes need to type in a URL, so try to minimize the risk of typos by using short names with all lower-case characters and no special characters (many people don't know how to type a ~).

5. Orphan Pages

Make sure that all pages include a clear indication of what web site they belong to since users may access pages directly without coming in through your home page. For the same reason, every page should have a link up to your home page as well as some indication of where they fit within the structure of your information space.

6. Long Scrolling Pages

Only 10% of users scroll beyond the information that is visible on the screen when a page comes up. All critical content and navigation options should be on the top part of the page.

Note added December 1997: More recent studies show that more people are scrolling. I still recommend minimizing scrolling on navigation pages, but it is no longer an absolute ban.

7. Lack of Navigation Support

Don't assume that users know as much about your site as you do. They always have difficulty finding information, so they need support in the form of a strong sense of structure and place. Start your design with a good understanding of the structure of the information space and communicate this structure explicitly to the user. Provide a site map and let users know where they are and where they can go.

8. Non-Standard Link Colors

Links to pages that have not been seen by the user are blue; links to previously seen pages are purple or red. Don't mess with these colors since the ability to understand what links have been followed is one of the few navigational aides that is standard in most web browsers. Consistency is key to teaching users what the link colors mean.

9. Outdated Information

Budget to hire a web gardener as part of your team. You need some-body to root out the weeds and replant the flowers as the website changes but most people would rather spend their time creating new content than on maintenance. In practice, maintenance is a cheap way of enhancing the con-tent on your website since many old pages keep their relevance and should be linked into the new pages. Of course, some pages are better off being removed completely from the server after their expiration date.

10. Overly Long Download Times

I am placing this issue last because most people already know about it; not because it is the least important. Traditional human factors guidelines indicate 10 seconds as the maximum response time before users lose inter-est. On the web, users have been trained to endure so much suffering that it may be acceptable to increase this limit to 15 seconds for a few pages.

For more links on useability visit http://www.useit.com/hotlist/

SOFTWARE: TRY BEFORE YOU BUY ON THE INTERNET

There are hundreds of sites devoted to software that might be of use to maintenance professionals. On the Internet, you can often try before you buy!

Since many retail outlets don't allow software returns unless the package is damaged, downloading from the Internet is a popular alternative. ComputerLife editors have selected their favorite download sites. ComputerLife's web site can be found at: www.computerlife.com

CAUTION

Caution is indicated whenever you download, because most viruses enter computer systems via the Internet through downloads or E-mail macros. The best action to take is before you start serously surfing. Purchase one of the major vendor's anti-virus program (good ones are marketed by Norton, McAfee) and keep it current. You keep an antivirus program up-to-date by visiting their web site and downloading their updates (usually you get a free 90 or 180 day subscription, after that time you have to pay by the year).

The sites listed below are swept for viruses regularly. A few might slip through. Be sure your anti-virus program running at all times. Some pro-grams allow you to scan incoming files. Again, viruses are being written and updated every day so keep your antivirus program up-to-date.

Favorite download sites for free software and shareware

1. ZDNet Software Library (http://www.hotfiles.com)

2. Shareware.com (http://www.shareware.com)

3. Download.com (http://www.download.com)
4. Windows95.com (http://www.windows95.com)
5. Happy Puppy (http://www.happypuppy.com)

For example I found an automobile maintenance package in the vertical shareware section of Download.com:

Auto-Do-It
From: **Champion Software Inc.**
Version: **2.1**
Date: **July 9, 1998**
File size: **2.3MB**
Category: **Vertical Markets**
Downloads: **1,394**
License: **Shareware**

Keep track of important vehicle maintenance information with Auto-Do-It. The program will document maintenance history and alert you of scheduled maintenance for multiple vehicles. In addition, users can print historical maintenance reports. Now, it includes a schedule reminder program that will notify you of upcoming maintenance from the icon in the Windows Taskbar.
Minimum requirements: **Windows 95/NT**

AUTOCAD

http://www.autodesk.com/ This is the URL for the company that owns AutoCAD. They provide training, user groups, and product support.

DISASTER RECOVERY

http://www.sgii.com/iw2 A new service available on the Internet is system backup of all your data files. This service allows you to keep a copy of critical files off-site and out of harm's way. They claim:

Unlimited
Off-site Backup and Storage
$9.95 a month

□ Works with AOL, CompuServe or any ISP
□ Never buy another Iomega Zip™ disk.
□ Totally confidential - military grade encryption.
□ The worlds most secure data center.

□   Installs in seconds - Windows/95 and NT 4.0.
□   Scheduler dials and backs up automatically.
□   Files are available 24 hours a day.
□   Retrieve any version of a file in seconds

http://www.fema.gov/cgi-shl/dbml.exe?action=query&template=/gems/g_index.dbm
Global Emergency Management System is a supersite for disaster issues operated by (FEMA Federal Emergency Management Agency **http://www.fema.gov**). The Global Emergency Management System (GEMS) is an online, searchable database containing links to Web sites in a variety of categories that are related in some way to emergency management. In most cases there is also a brief description of what the Website offers.

http://www.factorymutual.com The insurance giant Factory Mutual has a site for facility managers. This site has the information gained from 160 years of claims including how to assess a facilities' vulnerability and how to prepare against weather related disasters.

http://www.disaster.net Internet Disaster Information Center (IDIN) is provided as a public service by StarNet and Telekachina Productions, helps to distribute the latest news on disaster situations to the Internet community via the World Wide Web.

http://www.arkwright.com An insurance company. Their site has tips to reduce risks associated with disasters (both large and small).

QUALITY

The quality field is an active one on the Internet. There are newsgroups, supersites, consultants, and many published articles and books. Even a cursory search will yield useful resources.

http://www.dbainc.com/dba2/library/index.html Site courtesy of Dave Butler Assoc. has articles on TQM and services available.

INDUSTRIAL ENGINEERING

http://hbmaynard.com Meynard is one of the leading companies that supply productivity software and industrial engineering consultation. They offer training in work measurement, software to automate the mundane aspects of work measurement and support. Site is clear and loads quickly.

http://www.VisualT.com Simul8 is a simulation tool to help engineers predict where there will be problems with a new process and test fixes to known problems. The company is dedicated to teaching customers what simulation is, how to use it, and when it makes sense. They offer a simulation learning guide for download:

SIMULATION LEARNING GUIDE

What is simulation? How can I get results from simulation models? How can I build my own simulation models? What are the steps I have to take in a simulation study? This web based self-learning guide will tell you all you need to know to improve the operation that you manage using simulation.

LOGISTICS

http://www.logisticsweb.co.uk National Materials Handling Centre LogisticsWeb is a database of resources for the logistics discipline. It is located in the UK. Powerful site with links, publications, specialized search engine, breaking news, and job listings.

MAINTENANCE TRAINING

http://bama.ua.edu/~cstudies/ One of the most complete maintenance management training centers in the US. Its maintenance certificate program has graduated hundreds.

http://www.welding.org Hobart Institute of Welding Technology is a full-service-training institute for welding. They have videos, on-site training and classroom courses. Their web site had some welding puzzlers:

DO YOU KNOW YOUR WELDING HISTORY?

1. What ancient age of people used welding more than 2000 years ago?
2. When was the art of blacksmithing developed?
3. Who discovered acetylene?
4. What welding process was first used to weld railroad rails?
5. What year was automatic welding first introduced?
6. Gas metal arc welding was successfully developed in what year?

To find the answers, go to their web site (if they're not still up on the site, E-mail them for the answers!)

http://www.infraspection.com/ Infraspection Institute. Infrared educational certification (in several levels), educational resources, used equipment, jobs, stories, inspiring quotations, the whole nine yards.

http://www.willearn.com Williams Learning Network. Maintenance training services provides high-tech training in safety and maintenance for the shop floor.

http://www.wisc.edu/bschool/erdman The teaching field (especially at the college level) is changing. There are major shifts in marketing strategy among the universities. University of Wisconsin-Madison's School of Business has a Masters degree in Manufacturing and Technology Management. You can view the program at their site.

http://www.uophx.edu/online/ The URL for the on-line university within the University of Phoenix.

http://spider.rowan.edu/business/mi.htm Rowan University. Check out Rowan's certificate program for maintenance management. One of only three in the country.

PLANNING, SCHEDULING AND PROJECT MANAGEMENT

http://www.esi-intl.com ESI has their course "Managing Projects in Organizations," available on-line as well as other products

http://www.neosoft.com/~benchmrx/ An experimental site in planning and scheduling benchmarks. While most of the entries are from 1995-6, the site is useful for its discussions, links, and planning/scheduling forum. Their objectives are as follows:

OBJECTIVES

The number one purpose for publication of these problems is to accelerate the development and deployment of planning and scheduling technology that will have a significant impact in industry. In order to achieve this purpose, we have sought to satisfy the following objectives:

1. The problems must be relevant to industry. The emphasis will be on the ability to produce good solutions to realistic problems rather than on the ability to produce optimal solutions to abstract problems.

2. The problems must be realistic. They should be based upon or derived from real industrial planning and scheduling problems. They should not be made either easier or harder to accommodate the capabilities or goals of the research community.

3. The problems must advance the state of the practice. They should contain some features or attributes that challenge mainstream commercial products.

4. The problems must advance the state of the art. They should contain some features or attributes that challenge mainstream research results.

5. The problems must be public. They must be freely distributable and easily accessible over the Internet.

6. The problems must be open to all that wish to participate. Professors, graduate students, government and industrial laboratories, and commercial solution providers are all encouraged to participate.

7. The cost of entry should be low. The data should be easy to read and interpret. Participants should not need to invest a significant amount of effort in developing code to import and export problem data.

8. Practitioners at all levels of ability should be able to participate. The problems should be delivered in a graduated series so that participants can begin with simple cases and incrementally work their way to more challenging instances.

9. There should be some reward for participation. The solutions submitted by participants should be quickly and openly published so that all participants can easily compare their performance to others and can easily advertise their accomplishments.

10. There should be some mechanism for participants to publish comments on the problems sets, papers describing their own research, summaries of their products and services, and even problem sets of their own design.

---

http://www.teamflow.com CFM for process mapping and project management with downloadable evaluation copy.

http://www.salford.ac.uk/planning/ Planning and scheduling special interest group in the UK. Contains bibliographies, products, mailing lists, benchmarks, and links to other useful sites. While this is oriented to production, some of the theories apply in maintenance scheduling and job planning.

http://www.pmforum.org/ The home of the WWW Project Management Forum dedicated to international project management cooperation. Site includes resources such as book sales, journals, standards, vendor links, mailing list, SIGs.

SAFETY AND PPE (PERSONAL PROTECTIVE EQUIPMENT)

hhttp://www.cdc.gov/niosh/homepage.html National Institute of Occupation al Safety and Health This is an association for the promotion of safety in workplaces throughout the country. They offer research papers in many areas of industrial and public safety.

http://www.nsc.org National Safety Council. Their server houses the most complete library about safety: http://www.nsc.org/lrs/libtop.htm

http://www.retrieversoft.com Provides MSDS management software and a library of over 125,000 MSDS sheets. Software is also used as an automated maintenance and safety technical library.

http://www.safetycentral.org/isea Industrial Safety Equipment Association has several resources for safety managers. They maintain USA standards in personal protective equipment, have meetings, news, and provide buyer's guides and member services.

http://magicnet.net/~johnnw/ One of the great things about the Internet is that if someone is interested in a topic they can easily create a resource for others with similar interests. John W. created this site of resources for people interested in safety.

Another great aspect of the Internet is the ability to cross international lines. Staying in the health and safety field visit Great Britian for their site, **http://www.healthandsafety.co.uk.** You can tell from the uk domain that the site is housed in the UK. In New Zealand the health and safety office can be found at **http://osh.dol.govt.nz/index.html**. In neighboring Australia you can visit: **http://www.curtin.edu.au/curtin/dept/health/ohs/** Once again if you wanted to research health and safety for a new plant being built in Australia the Internet can help with a links page at **http://www.curtin.edu.au/curtin/dept/health/ohs/forum/links/htm**

Misc. Sites of Interest to maintenance professionals

http://www.tollfree.att.net/dir800 AT&T 800 directory

http://www.corrosion.com Data on corrosion and coatings.

http://www.industry.net/c/services/testmsr The Benchmarking Exchange is a resource of Industry Net. The site has resources for organizations wanting to benchmark their operation.

http://www.interchangeinc.com Parts Interchange Website.

http://www.pricewatch.com/ Price watch (computer prices).

http://www.iso14000.com/ ISO 14000 Information Center includes an overview of ISO 14000, a ISO 14000 guide, links and other organizations involved in the standard.

http://www.ansi.org/ American National Standards Institute is the center of standards activity in the USA (analogous to the ISO internationally).

http://www.truck.net/ Gives access to trucking companies, stores, magazines, DOT sites, trucking law, and trucking associations.

http://www.epa.gov/enviro/html/emci/chemref/index.html Database of toxic materials contains over 400 chemicals with complete information.

http://www_ivri.me.uic.edu/ The Virtual Reality Institute is dedicated to the development of virtual reality (VR) for industry. Site includes models, resources, and an ongoing project in VR.

http://www.industry.net/plcs/ A service of Industry Net called PLCs Online. Resources for those organizations involved in the use of PLCs.

Appendix1

History of the Internet

ROOTS OF THE TREE THE 1960s

In the 1960s, researchers began experimenting with linking computers to each other and to people through telephone hook-ups, using funds from the US Department of Defense Advanced Research Projects Agency (DARPA) is now known as ARPA. DARPA wanted to see if computers in different locations could be linked using packet switching a new technology which had the promise of letting several users share just one communications line. Previous computer networking efforts had required a line between each computer on the network, like a train track on which only one train can travel at a time.

The packet system allowed for creation of a data highway, in which large numbers of vehicles could essentially share the same lane. Each packet was given the computer equivalent of a map and a time stamp, so that it could be sent to the right destination, where it would then be reassembled into a message the computer or a human could use. Some highlights from the 1960s:

1962 Paul Baran, RAND: "On Distributed Communications Networks"
- ☐ Packet-switching (PS) networks; no single outage point

1965 ARPA sponsors study on "cooperative network of time-sharing computers"

1967 ACM Symposium on Operating Principles
- ☐ Plan presented for a packet-switching network
- ☐ First design paper on ARPANet published by Lawrence G. Roberts 1968 PS-network presented to the Advanced Research Projects Agency (ARPA) 1969 ARPANet commissioned by DoD

for research into networking

☐ First node at UCLA [Network Measurements Center - and soon after at:

☐ Stanford Research Institute (SRI) [SDS940/Genie], UCSB [Culler-Fried Interactive Mathematics - IBM 360/75:OS/MVT], U of Utah -[DEC PDP-10:Tenex],use of Information Message Processors (IMP) [Honeywell 516mini computer with 12K of memory] developed by Bolt Beranek and Newman, Inc. (BBN)

☐ U of Michigan, Michigan State and Wayne State U establish X.25-based Merit network for students, faculty, alumni

THE TREE BREAKS THROUGH INTO THE SUNLIGHT- THE 1970's

As we have seen, the Internet is a direct outgrowth this work by ARPA. Its function was to facilitate research communications with a network that could also keep military installations connected in the event of a nuclear strike or terrorist action.

In the 1970s, ARPA helped support the development of rules, or protocols, for transferring data between different types of computer networks. These "Internet" (from "internetworking") protocols made it possible to develop the worldwide Internet we have today. The name of the protocol that the entire Internet runs on is called TCP/IP (Transmission Control Program/ Internet Protocol).

By the close of the 1970s, links developed between ARPANet and counterparts in other countries. The world's computer networks were now being tied together. These first steps were like the beginning stringer that the spider shoots out to anchor the web she will build. In the 1970s major computing centers were starting to get tied together.

The state of the computer industry in the 1970s was very different from the industry today. The computer field was dominated by IBM with Univac, Control Data, Burroughs, Honeywell, and RCA sharing the small slice of the pie that was left. Upstarts like DEC were making a name in the minicomputer business and the first microcomputer kit was shipped in the mid 1970s. The first Apple computer came out in the late 1970s. The ARPANet reflected the industry alignments. Highlights from the 1970s:

1970 ALOHA Net at Univ. of Hawaii (connected to the ARPANet in 1972).

1971 15 nodes (23 hosts): UCLA, SRI, UCSB, U of Utah, BBN, MIT, RAND, SDC, Harvard, Lincoln Lab, Stanford, UIU(C), CWRU, CMU, and NASA/Ames

□ Ray Tomlinson of BBN invents email program to send messages across a distributed network.

□ International Conference on Computer Communications with demonstration of ARPANet between 40 machines organized by Bob Kahn.

□ InterNetworking Working Group (INWG) created to address need for establishing agreed upon protocols. Chairman: Vinton Cerf.

1973 First international connections to the ARPANet: University College of London (England) and Royal Radar Establishment (Norway)

□ Bob Kahn poses Internet problem, starts internetting research program at ARPA. Vinton Cerf sketches gateway architecture in March on back of envelope in hotel lobby in San Francisco

□ Cerf and Kahn present basic Internet ideas at INWG in September at U of Sussex, Brighton, UK

□ File Transfer Protocol (FTP) specification

1974 Vint Cerf and Bob Kahn publish "A Protocol for Packet Network Intercommunication" which specified in detail the design of a Transmission Control Program (TCP).

□ BBN opens Telenet, the first public packet data service (a commercial version of ARPANet)

1975 Operational management of Internet transferred to DCA (now DISA)

1976 Elizabeth II, Queen of the United Kingdom sends out an e-mail (various Net folks have e-mailed dates ranging from 1971 to 1978; 1976 was the most submitted and the only found in print)

□ UUCP (Unix-to-Unix copy) developed at AT&T Bell Labs and distributed with UNIX one year later.

1977 THEORYNET created by Larry Landweber at U of Wisconsin providing electronic mail to over 100 researchers in computer science (using a locally developed email system and TELENET for access to server).

1979 Meeting between U of Wisconsin, DARPA, NSF, and computer scientists from many universities to establish a Computer Science Department research computer network (organized by Larry Landweber).

□ USENET established using UUCP between Duke and UNC by Tom Truscott, Jim Ellis, and Steve Bellovin. All original groups were under net.* hierarchy.

THE TREE GROWS AND BUILDS A STRONG BASE- THE 1980'S

In the 1980s, this network of networks, which became known collectively as the Internet, expanded at a phenomenal rate. Hundreds, then thousands, of colleges, research companies and government agencies began to connect their computers to this worldwide network. Most of this growth was behind the scenes and invisible to the general public.

Some enterprising hobbyists and companies unwilling to pay the high costs of Internet access (or unable to meet stringent government regulations for access) learned how to link their own systems to the Internet, even if only for Email and conferences.

Some of these people formed companies that began offering access to the public. These enterprises became the ISPs (Internet Service Providers) of today. Now anybody with a computer and modem and persistence could tap into the world. Highlights of the 1980's:

1981 BITNET, the "Because It's Time NETwork"
- Started as a cooperative network at the City University of New York, with the first connection to Yale.
- Provides electronic mail and listserv (mailing list) servers to distribute information, as well as file transfers (FTP).
- CSNET (Computer Science NETwork) built by a collaboration of computer scientists and U. of Delaware, Purdue U., U. of Wisconsin, RAND Corporation and BBN through seed money granted by NSF to provide networking services (specially email) to university scientists with no access to ARPANet. CSNET later becomes known as the Computer and Science Network.

1982 DCA and ARPA establishes the Transmission Control Protocol (TCP) and Internet Protocol (IP), as the protocol suite, commonly known as TCP/IP, for ARPANet
- This leads to one of the first definitions of an "internet" as a connected set of networks, specifically those using TCP/IP, and "Internet" as connected TCP/IP internets.
- (Department of Defense declares TCP/IP suite to be standard for DoD networks
- Original connections between the Netherlands, Denmark, Sweden, and UK

1983 U of Wisconsin, no longer requiring users to know the exact path to other systems.
- ARPANet split into ARPANet and MILNET; the latter became intergrated withthe Defense Data Network created the previous year.

□ Desktop workstations come into being, many with Berkeley UNIX which includes IP (Internet Protocol) networking software.

1984 Domain Name Server (DNS) introduced.
□ Number of hosts breaks 1,000
□ Moderated newsgroups introduced on USENET (mod. *)

1985 100 years to the day of the last spike being driven on the cross-Canada railroad, the last Canadian university is connected to BITNET in a one year effort to have coast-to-coast connectivity.

1986 NSFNET (National Science Foundation Network) created the backbone speed of 56Kbps, with that is thousands of bits per second or about 6500 characters per second)
□ NSF establishes 5 super-computing centers to provide high-computing power for all (JVNC@Princeton, PSC@Pittsburgh, SDSC@UCSD, NCSA@UIUC, and Theory Center@Cornell).
□ This allows an explosion of connections, especially from universities.
□ NSF-funded SDSCNET, JVNCNET, SURANET, and NYSERNET operational

1987 NSF signs a cooperative agreement to manage the NSFNET backbone with Merit Network, Inc. (IBM and MCI involvement was through an agreement with Merit). Merit, IBM, and MCI later founded ANS.
□ Email link established between Germany and China using CSNET proto cols, with the first message from China sent on 20 September.
□ Number of hosts breaks 10,000
□ Number of BITNET hosts breaks 1,000

1988 1 November - Internet worm (Internet computer virus) burrows through the Net, affecting ~6,000 of the 60,000 hosts on the Internet.
□ CERT (Computer Emergency Response Team) formed by DARPA in response to the needs exhibited during the Morris worm incident. The worm is the only advisory issued this year.
□ NSFNET backbone upgraded to T1 (1.544Mbps that's 1,544,000 bits per second or about 200,000 characters per second)
□ Internet Relay Chat (IRC) developed by Jarkko Oikarinen
□ Countries connecting to NSFNET: Canada, Denmark, Finland, France, Iceland, Norway, Sweden

1989 # of hosts breaks 100,000

- ☐ First relays between a commercial electronic mail carrier and the Internet: MCI Mail through the Corporation for the National Research Initiative (CNRI), and CompuServe through Ohio State University

- ☐ Cuckoo's Egg written by Clifford Stoll tells the real-life tale of a German cracker group who infiltrated numerous US facilities

- ☐ CERT (Computer Emergency Response Team) advisories: 7

- ☐ Countries connecting to NSFNET: Australia, Germany, Israel, Italy, Japan, Mexico, Netherlands, New Zealand, Puerto Rico, United Kingdom

THE 1990S THE TREE FLOURISHES

President Clinton and Vice President Gore start to compare the Information Super Highway to the Interstate Highway system in size and scope. The Internet took off in the popular press in 1994. The biggest change was the World Wide Web (WWW) invented by some programmers at CERN in Switzerland. Software browsers became available that allowed users to access pictures, color, animation, and video. By 1995 hundreds, then thousands then hundreds of thousands of sites were created on the WWW that could be visited. At the end of 1997 there were eighty million web sites.

Some estimates are that the volume of messages transferred through the Net grows 20 percent a month. In response, government and other users have tried in recent years to expand the Net itself. Once, the main Net backbone in the United States moved data at 56,000 bits per second (1986), then 1.5 million bits per second (1988). These speeds proved too slow for the ever-increasing amounts of data being sent over it, and in the early 1990s the maximum speed was increased to 45 million bits per second (1991).

Even before the Net was able to reach that speed, however, Net experts were already figuring out ways to pump data at speeds of up to 2 billion bits per second, fast enough to send the entire Encyclopedia Britannica across the country in just one or two seconds. Another major change has been the development of commercial services that provide Internetworking services at speeds comparable to those of the government system. Highlights from the 1990s:

1990 ARPANet ceases to exist

- ☐ Archie released by Peter Deutsch, Alan Emtage, and Bill Heelan at McGill

- ☐ CAnet formed by 10 regional networks as national Canadian backbone with direct connection to NSFNET

- The first remotely operated machine to be hooked up to the Internet, the Internet Toaster, (controlled via SNMP) makes its debut at Interop.
- CERT advisories: 12, reports: 130
- Countries connecting to NSFNET: Argentina, Austria, Belgium, Brazil, Chile, Greece, India, Ireland, Korea, Spain, Switzerland

1991 Wide Area Information Servers (WAIS), invented by Brewster Kahle, released by Thinking Machines Corporation

- Gopher released by Paul Lindner and Mark P. McCahill from the U of Minn
- World-Wide Web (WWW) released by CERN; Tim Berners-Lee developer
- US High Performance Computing Act (Gore 1) establishes the National Research and Education Network (NREN)
- NSFNET backbone upgraded to T3 (44.736Mbps)
- NSFNET traffic passes 1 trillion bytes/month and 10 billion packets/month
- CERT advisories: 23
- Countries connecting to NSFNET: Croatia, Czech Republic, Hong Kong, Hungary, Poland, Portugal, Singapore, South Africa, Taiwan, Tunisia

1992 Internet Society (ISOC) is chartered

- Number of hosts breaks 1,000,000
- Veronica, a gopherspace search tool, is released by University of Nevada
- World Bank comes on-line
- Japan's first ISP, Internet Initiative Japan (IIJ), is formed by Koichi Suzuki
- The term "Surfing the Internet" is coined by Jean Armour Polly
- CERT advisories: 21, reports: 800
- Countries connecting to NSFNET: Cameroon, Cyprus, Ecuador, Estonia, Kuwait, Latvia, Luxembourg, Malaysia, Slovakia, Slovenia, Thailand, Venezuela

1993 InterNIC created by NSF to provide specific Internet management services: directory and database services (AT&T), registration services (Network Solutions Inc.)

- US White House comes on-line (http://www.whitehouse.gov/):
- President Bill Clinton: president@whitehouse.gov
- Vice-President Al Gore: vice-president@whitehouse.gov
- First Lady Hillary Clinton: @whitehouse.gov
- Internet Talk Radio begins broadcasting

- □ United Nations (UN) come on-line
- □ US National Information Infrastructure Act, Businesses and media really take notice of the Internet
- □ Mosaic takes the Internet by storm; WWW proliferates at a 341,634% annual growth rate of service traffic. Gopher's growth is 997%.
- □ CERT advisories: 18, reports: 1300
- □ Countries connecting to NSFNET: Bulgaria, Costa Rica, Egypt, Fiji, Ghana, Guam, Indonesia, Kazakhstan, Kenya, Liechtenstein, Peru, Romania, Russian Federation, Turkey, Ukraine, UAE, US Virgin Islands

1994 ARPANet/Internet celebrates 25th anniversary
- □ Communities begin to be wired up directly to the Internet (Lexington and Cambridge, Mass., USA)
- □ Shopping malls arrive on the Internet
- □ Arizona law firm of Canter & Siegel "spams" the Internet with email advertising green card lottery services; Net citizens flame back
- □ NSFNET traffic passes 10 trillion bytes/month
- □ Yes, it's true - you can now order pizza from the Hut online
- □ WWW edges out telnet to become 2nd most popular service on the Net (behind ftp-data) based on % of packets and bytes traffic distribution on NSFNET
- □ Japanese Prime Minister on-line (http://www.kantei.go.jp/)
- □ First Virtual, the first cyberbank, open up for business
- □ CERT advisories: 15, reports: 2300
- □ Countries connecting to NSFNET: Algeria, Armenia, Bermuda, Burkina Faso, China, Colombia, French Polynesia, Jamaica, Lebanon, Lithuania, Macao, Morocco, New Caledonia, Nicaragua, Niger, Panama, Philippines, Senegal, Sri Lanka, Swaziland, Uruguay, Uzbekistan

1995 NSFNET reverts back to a research network. Main US backbone traffic now routed through interconnected network providers
- □ The new NSFNET is born as NSF establishes the very high speed Backbone Network Service (vBNS) linking super-computing centers: NCAR, NCSA, SDSC, CTC, PSC
- □ Hong Kong police disconnect all but 1 of the colony's Internet providers in search of a hacker. 10,000 people are left without Net access.
- □ RealAudio, an audio streaming technology, lets the Net hear in near realtime

- Radio HK, the first 24 hr., Internet-only radio station starts broadcasting

- WWW surpasses ftp-data in March as the service with greatest traffic on NSFNet based on packet count, and in April based on byte count

- Traditional online dial-up systems (CompuServe, America Online, Prodigy) begin to provide Internet access

- A number of Net related companies go public, with Netscape leading the pack with the 3rd largest ever NASDAQ IPO share value (9 August)

- Registration of domain names is no longer free. Beginning 14 September, a $50 annual fee has been imposed, which up until now was subsidized by NSF. NSF continues to pay for .edu registration, and on an interim basis for .gov

- The Vatican comes on-line (http://www.vatican.va/)

- The Canadian Government comes on-line (http://canada.gc.ca/)

- The first official Internet wiretap was successful in helping the Secret Service and Drug Enforcement Agency (DEA) apprehend three individuals who were illegally manufacturing and selling cell phone cloning equipment and electronic devices

- Operation Home Front connects, for the first time, soldiers in the field with their families back home via the Internet.

- CERT advisories: 18, reports: 2412

- Country domains registered: Ethiopia (ET), Cote d'Ivoire (CI), Cook Islands (CK) Cayman Islands (KY), Anguilla (AI), Gibraltar (GI), Vatican (VA), Kiribati (KI), Kyrgyzstan (KG), Madagascar (MG), Mauritius (MU), Micronesia (FM), Monaco (MC), Mongolia (MN), Nepal (NP), Nigeria (NG), Western Samoa (WS), San Marino (SM), Tanzania (TZ), Tonga (TO), Uganda (UG), Vanuatu (VU)

- Technologies of the Year: WWW, Search engines
  1996 Internet phones catch the attention of US telecommunication com panies who ask the US Congress to ban the technology (which has been around for years)

- The controversial US Communications Decency Act (CDA) be-comes law in the US in order to prohibit distribution of indecent materials over the Net. A few months later a three-judge panel imposes an injunction against its enforcement. Supreme Court rules most of it unconstitutional in 1997.

- 9,272 organizations find themselves unlisted after the InterNIC drops their name service as a result of not having paid their domain name fee

- Various ISPs suffer extended service outages, bringing into question whether they will be able to handle the growing number of users. AOL (19 hours), Netcom (13 hours), AT&T WorldNet (28 hours - email only)

- MCI upgrades Internet backbone adding ~13,000 ports, bringing the effective speed from 155Mbps to 622Mbps.

☐ The Internet Ad Hoc Committee announces plans to add 7 new generic Top Level Domains: .firm, .store, .web, .arts, .rec, .info, .nom. The IAHC plan also calls for a competing group of domain registrars worldwide.

☐ A cancelbot (program that erases messages to combat spamming) is released on USENET wiping out more than 25,000 messages.

☐ The WWW browser war, fought primarily between Netscape and Microsoft, has rushed in a new age in software development, whereby new releases are made quarterly with the help of Internet users eager to test upcoming (beta) versions.

☐ Restrictions on Internet use around the world:

China: requires users and ISPs to register with the police

Germany: cuts off access to some newsgroups carried on CompuServe

Saudi Arabia: confines Internet access to universities and hospitals

Singapore: requires political and religious content providers to register with the state

New Zealand: classifies computer disks as "publications" that can be censored and seized

Source: Human Rights Watch

☐ vBNS additions: Baylor College of Medicine, Georgia Tech, Iowa State Univ., Ohio State Univ., Old Dominion Univ., Univ. of CA, Univ. of CO, Univ. of Chicago, Univ. of IL, Univ. of MN, Univ. of PA, Univ. of TX, Rice Univ.

☐ CERT advisories: 27, reports: 2573

☐ Country domains registered: Qatar (QA), Central African Republic (CF), Mauritania (MF), Oman (OM), Norfolk Island (NF), Tuvalu (TV), French Polynesia (PF), Syria (SY), Aruba (AW), Cambodia (KH), French Guyana (GF), Eritrea (ER), Cape Verde (CV), Burundi (BI), Benin (BJ) Bosnia-Herzegovina (BA), Andorra (AD), Guadeloupe (GP), Guernsey (GG), Isle of Man (IM), Jersey (JE), Lao (LA), Maldives (MV), Marshall Islands (MH), Mauritania (MR), Northern Mariana Islands (MP), Rwanda (RW), Togo (TG), Yemen (YE), Zaire (ZR)

☐ Technologies of the Year: Search engines, JAVA, Internet Phone

1997 2000th RFC: "Internet Official Protocol Standards"

☐ 71,618 mailing lists registered at Liszt

☐ The American Registry for Internet Numbers (ARIN) is established to handle administration and registration of IP numbers to the geographical areas currently handled by Network Solutions (InterNIC), starting March 1998.

☐ Country domains registered: Falkland Islands (FK), East Timor (TP), Congo (CG), Christmas Island (CX), Gambia (GM), Guinea-Bissau (GW), Haiti (HT), Iraq (IQ), Libya (LY), Malawi (MW), Martinique (MQ), Montserrat (MS), Myanmar (MM), French Reunion Island (RE), Seychelles (SC), Sierra Leone (SL), Sudan (SD), Turkmenistan (TM), Turks and Caicos Islands (TC)

The timeline is reprinted with permission from Robert H'obbes' Zakon, who can be reached at E-mail address: zakon@info.isoc.org .

The whole report is archived at:
http://info.isoc.org/guest/zakon/Internet/History/HIT.html

The above history section was also partially adapted from work by Adam Gaffin (adamg@world.std.com) Senior Reporter, Middlesex News, Farmingham, MA.

 *Appendix 2*

ALPHABETICAL LIST OF WEB SITES

A

Academy of Infrared Thermography http://academy-of-infrared.net
Advanced Maintenance Solutions http://cogz.com/cmmsa.htm
Advanced Software Design http://www.asd-info.com/eval.htm
Airplane Maintenance http://www.amtonline.com/
AltaVista http://www.Altavista.digital.com
Americans with Disabilities Act Site http://www.usdoj.gov/crt/ada/adahom1.htm
American National Standards Institute http://www.ansi.org/
Architecture, Engineering, and Construction (Canada)
          http://ctca.unb.ca/CTCA/sources/
Arkwright http://www.arkwright.com
ASME http://mecheng.asme.org
Amazon.com http://www.amazon.com
American Society for Industrial Security http://www.asisonline.org
AT&T 800 directory http://www.tollfree.att.net/dir800
AutoCAD http://www.autodesk.com/
AutomationNET http://www.automationnet.com

B

Benchmate Systems http://www.benchmate.com
BMC http://www.ultranet.ca/bmc/bmc.htm
Burke & Associates http://burkesystems.com/Software.htm
Dave Butler Assoc. http://www.dbainc.com/dba2/library/index.html

C

Cat Pump http://www.usinternet.com/catpumps
CFM http://www.teamflow.com
Cleaning Management http://www.cleannet.com
Climax Machine Tools http://www.cpmt.com
Consulting-Specifying Engineer magazine http://www.csemag.com
Curtin University (Australia) program in occupational safety and health
          http://www.curtin.edu.au/curtin/dept/health/ohs/

Curtin University Health and Safety links
http://www.curtin.edu.au/curtin/dept/health/ohs/forum/links/htm
Cybertown http://www.cybertown.com/campeng.html

D

Darex http://darex.com/sharpeners
DARPA (Defense Advanced Research Projects Agency) http://www.arpa.mil
Datastream http://www.dstm.com
DejaNews http://www.dejanews.com/
Department of Agriculture http://www.usda.gov/
Department of Commerce http://www.doc.gov
Department of Education http://www.ed.gov/
Department of Energy Grants http://apollo.asti.gov/home.html
Department of Justice http://www.usdoj.gov
Department of Labor http://www.dol.gov
Department of Transportation http://www.dot.gov
Design Info http://www.designinfo.com
Desktop Innovations http://mainboss.com
DHR Maintenance engineering http://www.dhr.com
Dingo Maintenance Systems http://dingos.com
Disaster recovery http://www.sgii.com/iw2
DMSI http://www.desmaint.com/
DOD Safety site http://www.acq.osd.mil/ens/sh/
DOD Aircraft Safety http://www.acq.osd.mil/ens/sh/faa.gif
DOD Virtual Library (procurement)
http://www.arnet.gov/References/References.html
Dogpile www.dogpile.com
Download.com http://www.download.com
DP Solutions Inc. http://www.dpsi-cmms.com/

E

Eagle Technology http://www.eagleone.com
E-commerce at DOD http://www.acq.osd.mil/ec
Entek http://www.entek.com
EPA http://epa.gov
EPA Energy Star program http://epa.gov/docs/GCDOAR/energystar.html
EPA Database of Toxic Materials
http://www.epa.gov/enviro/html/emci/chemref/index.html
ESI (project management) http://www.esi-intl.com
Excite http://www.Excite.com

F

Factory Mutual http://www.factorymutual.com
Federal Emergency Management Agency http://www.fema.gov/
Fed World http://www.fedworld.gov/
Fed World Government jobs http://www.fedworld.gov/jobs/jobsearch.html
Food and Drug Administration http://www.fda.gov
Findlaw http://www.findlaw.com
Fleet Owner http://www.fleetowner.com/default.html

Framatome Technologies http://www.framatech.com/marketing/Empath.asp
Frank Measier http://www.efn.org/~franka/MMeasier/TIPS.html

G

Galaxy Search http://galaxy.tradewave.com/search.html
General Electric http://www.ge.com/edc/index.html
General Services Administration http://www.gsa.gov.
General Pump http://www.generalpump.com
Global Emergency Management System (GEMS)
   http://www.fema.gov/cgishl/dbml.exe?action=query&template=/gems/g_index.dbm
W.W. Grainger http://www.grainger.com

H

Happy Puppy http://www.happypuppy.com
Health and Safety UK http://www.healthandsafety.co.uk
Health and Safety New Zealand http://osh.dol.govt.nz/index.html
Hercules Chemical http://www.herchem.com/
Hobart Institute of Welding http://www.welding.org
Honeywell http://www.iac.honeywell.com/
Hotbot http://www.Hotbot.com
HUD http://www.hud.gov

I

Industry Net Benchmarking Exchange http://www.industry.net/c/services/testmsr
Industry Net (PLCs Online) http://www.industry.net/plcs
Industrial Safety Equipment Association http://www.safetycentral.org/isea
Inframetrics http://www.inframetrics.com
Infraspection Institute http://www.infraspection.com/
Internet Disaster Information Center (IDIN) http://www.disaster.net
ISO 14000 http://www.iso14000.com/
IMPRO http://www.impomag.com/
Industrial Link http://www.industrylink.com/
Industrial Press http://www.industrialpress.com
Industrial Sourcebook http://www.industrialcourcebook.com
Industry Net http://www.industrynet.com
Industrial View http://www.industrialview.com
The Industrial Distribution Association http://www.industry.net/c/orgindex/ida
Iron and Steel Institute http://www.steel.org/
Industrial trade journals (site by AT&T) http://www.ichange.com
Infoseek http://www.Infoseek.com

J

James S. Cullen http://home.fuse.net/MMCS/
JB Systems http://www.jbsystems.com
Johnson Controls http://www.johnsoncontrols.com/cg
John W's. safety site http://magicnet.net/~johnnw/

K

The Kinsey group http://www.thekinseygroup.com/bck-grnd.htm

L

Liberty Technologies http://www.libertytech.com
Library of Congress http://www.loc.gov/
Life Cycle Engineering http://www.lce.com/index.html
Logistics and warehousing magazines (listing) can be found at
        http://www.cap-ai.com/library2/magazin.html
LogisticsWeb National Materials Handling Center http://www.logisticsweb.co.uk
Lycos http://www.Lycos.com

M

Macmillan Press http://www.mcp.com/
Maintenance Management book list in the UK
http://www.conference-communication.co.uk/books/Default.htm
Management Technology Inc http://www.managementtechnologies.com
Manufacturing Online (http://www.chesapk.com/~chesapeake/
Manufacturing and the Internet http://mtiac.hq.iitri.com/mtiac/mfg.htm
Manufacturing Information Net http://mfginfo.com/home.htm
Manufacturing Information Solutions Web site http://conduit2.sils.umich.edu/
Mapquest http://www.mapquest.com
Marcam Corp. http://www.avantis.marcam.com
Max Hon http://www.maxhon.com.hk/index1.htm
McGraw Hill http://www.mcgraw-hill.com/books.html
McMaster-Carr http://www.mcmastercarr.com
Mechanical Engineering Magazine http://www.memagazine.org/index.html
Metal Machining and Fabrication International Internet Directory
        http://www.mmf.com/metal/
Meynard http://hbmaynard.com
Microwest http://www.microwst.com
Mining Internet service http://www.miningusa.com
Momma-Mother of all search engines www.mamma.com
Motion Industries http://www.motion-industries.com
Multi-Media Handbook for Engineering Design
        http://www.dig.bris.ac.uk/hbook
Multi search engine http://m5.interence.com/ifind/

N

NASA http://hypatia.gsfc.nasa.gov/NASA_homepage.html
NASA abstract and technical report services
        http://techreports.larc.nasa.gov/cgi-bin/ntrs
National Information Center for Reliability Engineering, University of Maryland
        www.glue.umd.edu/enre/reinfo.htm
        Toolspage: http://www.enre.umd.edu/rmp.htm
National Institute for Occupational Safety and Health
        http://www.cdc.gov/niosh/homepage.html
National Institute of Standards and Technology http://www.nist.gov/welcome.html
National Institute of Standards and Technology Standard Reference Database
        http://www.nist.gov/srd/srd.htm
National Safety Council http://www.nsc.org
National Safety Council Safety Library http://www.nsc.org/lrs/libtop.htm

National Science Foundation http://www.nsf.gov/home/pubinfo/start.htm
Naval Research Laboratory http://www.nrl.navy.mil/
NAVFAC http://www.navy.mil/homepages/navfac
NCEF (National Clearinghouse on Educational Facilities)
        http://www.edfacilities.org/index.html
NCEF Supersite http://www.edfacilities.org/links.html
New Equipment Digest http:// www.newequipment.com.
NIST (National Institute of Standards) http://www.nist.gov
NIST (manufacturing engineering laboratory) http://nist.gov/mel/melhome.html
Norwich Technologies http://www.norwichtech.com
NTIS http://www.ntis.gov/databases/armypub.htm

O

Omega http://www.omega.com
Omnicomp (subsidiary of Enron Corp) http://www.omni-comp.com
Open Text Index http://index.opentext.net/
OSHA http://www.osha.gov/

P

Part Net Http://www.Part.net/
Paul Reichert's http://weber.u.washington.edu/~brennon/reichert/profile.html
Peterson Alignment Tools http://www.petersontools.com
Planning and scheduling benchmarks http://www.neosoft.com/~benchmrx/
Planning and scheduling SIG (UK) http://www.salford.ac.uk/planning/
Polibrid (corrosion and coatings).http://www.corrosion.com
Postal Service http://www.usps.gov/
Power Transmission Handbook (excerpts) http://www.hoppenstedt.de/componet
Precision Micro Dynamics http://www.Pmdi.com/php2.php
Price watch (computer prices)http://www.pricewatch.com/
Process Pumps and Filtration http://www.iglou.com/pitt
Project Management Forum http://www.pmforum.org
Projetech consultants http://www.projetech.com
PSDI http://www.psdi.com
Plant Maintenance Resource Center
        http://www.iinet.net.au/~sdunn/maintenance/CMMS_vendors.html
Plant Services http://www.plantservices.com
Popular Mechanics http://popularmechanics.com/
Prentice Hall http://www.prenhall.com/
Prism Computer http://www.prismcc.com/famis
Pruftechnik http://www.pruftechnik.com
PSDI http://www.psdi.com

R

Redlake Camera http://www.redlake.com/imaging
Reliability http://reliability.com
Reliability Analysis Center http://rome.iitri.com/rac/
Reed Exposition http://www.techexpo.com
Remstar http://www.remstar.com
RetrieverSoft http://www.retrieversoft.com

Revere, Inc http://www.walker.com
Rose-Hulman Institute of Technology
http://www.civeng.carleton.ca/Other_Info/Other_Information.html
Roy Jorgenson http://www.royjorgensen.com/home/index.html

S

SavvySearch http://www.cs.colostate.edu/~dreiling/smartform.html
Sensors magazine http://www.sensorsmag.com/
Shareware.com (http://www.shareware.com)
Shock and Vibration Analysis Center http://saviac.usae.bah.com
Simul8 http://www.VisualT.com
SISI Group, Inc. http://www.isigroup.com
SKF   http://www.skf.com
SKF (internal magazine) http://evolution.skf.com/gb/eng-main.asp
Small Business Admin http://www.sbaonline.sba.gov/
Sofwave Maintenance Information Systems http://www.sofwave.com
Somax http://www.somax.com
Springfield Resources http://www.maintrainer.com
Stanford University Mechanical Engineering Library
http://cdr.stanford.edu/html/WWW-ME/home.html
Stanley Proctor http://www.stanleyproctor.com
Sunbelt Engineering http://www.sunbeltengineering.com/mis_lnk.htm
Sweets http://www.sweets.com/

T

TechLabs http://www.techlabs.com/techlabs/ - techlabs
Thomas Register http://www.thomasregister.com.
TMA http://www.tmasys.com
Today's Facility Manager http://www.tfmgr.com
Trademark and Patent office http://www.uspto.gov
Tribology http://www.shef.ac.uk/~mpe/mattrib/tribology
Truck Net http://www.truck.net/
TSW International http://www.tswi.com
Trader on-line http://www.traderonline.com/equip/index.shtml
Trane http://www.trane.com
Truck Fleet Magazine http://www.truckfleetmgtmag.com/
Turbo Guide . http://www.turboguide.com/
TWI Press http://www.twi-press.com

U

UE Systems http://www.uesystems.com
University of California -Davis Advanced Highway Maintenance
http://www-ahmct.engr.ucdavis.edu/ahmct/
University of Massachusetts library
http://www.ecs.umass.edu/mie/labs/mda/dlib/dlib.html
University of Phoenix on-line http://www.uophx.edu/online/
University of Tennessee Maintenance Technology Laboratory
http://www.engr.utk.edu/mrc/

University of Toronto (Business Process Reengineering)
    http://www.ie.utoronto.ca/EIL/tool/BPR.html
University of Wisconsin-Madison School of Business
    http://www.wisc.edu/bschool/erdman

V

VibrAlign http://www.vibralign.com
Vibration Institute of Canada http://www.vibrate.net
Virtual Reality Institute http://www_ivri.me.uic.edu/

WXZ

Williams Learning Network http://www.willearn.com
Weather forecast http://cirrus.sprl.umich.edu/wxnet/
Web-based information and research resource http://www.eiq.com
Web Crawler http://www.Webcrawler.com
Western Software http://www.westernsoftware.com/products.htm
Williams Learning Network http://www.willearn.com
Windows95.com http://www.windows95.com
WinterCress http://www.aa.net/~winter
WWW virtual library (Engineering section)
    http://www.w3.org/hypertext/DataSources/bySubject/Overview.html
WWW virtual library on energy
    http://solstice.crest.org/online/virtual-library/Vlib-energy.html
Yahoo http://www.yahoo.com
ZDNet Software Library http://www.hotfiles.com

Appendix 3

LIST OF WEB SITES BY CATAGORY

ASSOCIATION SITES
Academy of Infrared Thermography http://academy-of-infrared.net
American National Standards Institute http://www.ansi.org/
American Society for Industrial Security http://www.asisonline.org
American Society of Mechanical Engineers (ASME)
        http://www-jmd.engr.ucdavis.edu/jmd/
Industrial Safety Equipment Association http://www.safetycentral.org/isea
Iron and Steel Institute http://www.steel.org/
Metal Machining and Fabrication International Internet Directory
        http://www.mmf.com/metal/
National Institute for Occupational Safety and Health
        http://www.cdc.gov/niosh/homepage.html
National Safety Council http://www.nsc.org.
National Safety Council Safety Library http://www.nsc.org/lrs/libtop.htm
Vibration Institute of Canada http://www.vibrate.net

CMMS AND PDM VENDORS
Academy of Infrared Thermography http://academy-of-infrared.net
Advanced Maintenance Solutions http://cogz.com/cmmsa.htm
Advanced Software Design http://www.asd-info.com/eval.htm
Benchmate Systems http://www.benchmate.com
Burke & Associates http://burkesystems.com/Software.htm
Datastream http://www.dstm.com
Desktop Innovations http://mainboss.com
Dingo Maintenance Systems http://dingos.com
DMSI http://www.desmaint.com/
DP Solutions Inc. http://www.dpsi-cmms.com/
Eagle Technology http://www.eagleone.com
Entek http://www.entek.com
Framatome Technologies http://www.framatech.com/marketing/Empath.asp
Inframetrics http://www.inframetrics.com

JB Systems http://www.jbsystems.com
Liberty Technologies http://www.libertytech.com
Marcam Corp. http://www.avantis.marcam.com
Microwest http://www.microwst.com
Norwich Technologies http://www.norwichtech.com
Omnicomp (subsidiary of Enron Corp) http://www.omni-comp.com
Prism Computer http://www.prismcc.com/famis
PdMA http://www.pdma.com
Peterson Alignment Tools http://www.petersontools.com
Pruftechnik http://www.pruftechnik.com
PSDI http://www.psdi.com
Redlake Camera http://www.redlake.com/imaging
Reliability http://reliability.com
Revere, Inc http://www.walker.com
SISI Group, Inc. http://www.isigroup.com
Sofwave Maintenance Information Systems http://www.sofwave.com
Somax http://www.somax.com
TMA http://www.tmasys.com
Tribology http://www.shef.ac.uk/~mpe/mattrib/tribology
TSW International http://www.tswi.com
Turbo Guide http://www.turboguide.com/
UE Systems http://www.uesystems.com
VibrAlign http://www.vibralign.com
Western Software http://www.westernsoftware.com/products.htm
WinterCress http://www.aa.net/~winter

GOVERNMENT SITES

Americans with Disabilities Act Site http://www.usdoj.gov/crt/ada/adahom1.htm
DARPA (Defense Advanced Research Projects Agency) http://www.arpa.mil
DOD Safety site http://www.acq.osd.mil/ens/sh/
DOD Aircraft Safety http://www.acq.osd.mil/ens/sh/faa.gif
DOD Virtual Library(procurement)
        http://www.arnet.gov/References/References.html
Department of Agriculture http://www.usda.gov/
Department of Commerce http://www.doc.gov:
Department of Education http://www.ed.gov/
Department of Energy Grants http://apollo.asti.gov/home.html.
Department of Justice http://www.usdoj.gov
Department of Labor http://www.dol.gov
Department of Transportation http://www.dot.gov
E-commerce at DOD http://www.acq.osd.mil/ec
EPA http://epa.gov
EPA Energy Star program http://epa.gov/docs/GCDOAR/energystar.html
EPA Database of Toxic Materials
        http://www.epa.gov/enviro/html/emci/chemref/index.html
Federal Emergency Management Agency http://www.fema.gov/
Fed World http://www.fedworld.gov/
Fed World Government jobs http://www.fedworld.gov/jobs/jobsearch.html
Food and Drug Administration http://www.fda.gov

General Services Administration http://www.gsa.gov
Global Emergency Management System
    http://www.fema.gov/cgishl/dbml.exe?action=query&template=/gems/g_index.dbm
HUD http://www.hud.gov
Library of Congress http://www.loc.gov/.
NASA http://hypatia.gsfc.nasa.gov/NASA_homepage.html
NASA abstract and technical report services.
    http://techreports.larc.nasa.gov/cgi-bin/ntrs
National Science Foundation http://www.nsf.gov/home/pubinfo/start.htm
National Institute of Standards and Technology http://www.nist.gov/welcome.html
National Institute of Standards and Technology Standard Reference Database
    http://www.nist.gov/srd/srd.htm
Naval Research Laboratory http://www.nrl.navy.mil/
NAVFAC http://www.navy.mil/homepages/navfac
NCEF (National Clearinghouse on Educational Facilities)
    http://www.edfacilities.org/index.html
NCEF Supersite http://www.edfacilities.org/links.html
NIST (National Institute of Standards) http://www.nist.gov
NIST (manufacturing engineering laboratory) http://nist.gov/mel/melhome.html
NTIS http://www.ntis.gov/databases/armypub.htm
Postal Service http://www.usps.gov/
OSHA http://www.osha.gov/
Small Business Admin http://www.sbaonline.sba.gov/
Trademark and Patent office http://www.uspto.gov

INDUSTRIAL DISTRIBUTOR

W.W. Grainger http://www.grainger.com
McMaster-Carr http://www.mcmastercarr.com
Motion Industries http://www.motion-industries.com
The Industrial Distribution Association http://www.industry.net/c/orgindex/ida

EQUIPMENT AND COMPONENT SUPPLIERS

Cat Pump http://www.usinternet.com/catpumps
Darex http://darex.com/sharpeners
Climax Machine Tools http://www.cpmt.com
General Electric http://www.ge.com/edc/index.html
GE Capital and GE Rental http://www.gecsn.com/
General Pump http://www.generalpump.com
Hercules Chemical http://www.herchem.com/
Honeywell http://www.iac.honeywell.com/
Industrial Sourcebook http://www.industrialcourcebook.com
Johnson Controls http://www.johnsoncontrols.com/cg
Machine Tools http://machinetools.com/
Omega http://www.omega.com
Part Net Http://www.Part.net/
Precision Micro Dynamics http://www.Pmdi.com/php2.php
Remstar http://www.remstar.com
SKF   http://www.skf.com
Stanley Proctor http://www.stanleyproctor.com

Trader on-line http://www.traderonline.com/equip/index.shtml
Trane http://www.trane.com

MAINTENANCE MAGAZINES
Airplane Maintenance http://www.amtonline.com/
Cleaning Management http://www.cleannet.com
Consulting-Specifying Engineer magazine http://www.csemag.com.
Fleet Owner http://www.fleetowner.com/default.html
IMPRO http://www.impomag.com/
Logistics and warehousing magazines (listing) can be found at
http://www.cap-ai.com/library2/magazin.html
Mechanical Engineering Magazine http://www.memagazine.org/index.html
New Equipment Digest http://www.newequipment.com.
Plant Services http://www.plantservices.com
Popular Mechanics http://popularmechanics.com/
Process Pumps and Filtration http://www.iglou.com/pitt
Sensors magazine http://www.sensorsmag.com/
SKF (internal magazine) http://evolution.skf.com/gb/eng-main.asp
Today's Facility Manager http://www.tfmgr.com
Truck Fleet Magazine http://www.truckfleetmgtmag.com/

MAINTENANCE CONSULTANTS
BMC http://www.ultranet.ca/bmc/bmc.htm
DHR Maintenance engineering http://www.dhr.com
James S. Cullen http://home.fuse.net/MMCS/
The Kinsey group http://www.thekinseygroup.com/bck-grnd.htm
Life Cycle Engineering http://www.lce.com/index.html
Management Technology Inc http://www.managementtechnologies.com
Max Hon http://www.maxhon.com.hk/index1.htm
Projetech consultants http://www.projetech.com
Paul Reichert's http://weber.u.washington.edu/~brennon/reichert/profile.html
Reliability http://reliability.com
Roy Jorgenson http://www.royjorgensen.com/home/index.html
Springfield resources http://www.maintrainer.com
Sunbelt Engineering http://www.sunbeltengineering.com/mis_lnk.htm
TechLabs http://www.techlabs.com/techlabs/ - techlabs

MISCELLANEOUS SITES
Arkwright http://www.arkwright.com
AT&T 800 directory http://www.tollfree.att.net/dir800
AutoCAD http://www.autodesk.com/
CFM http://www.teamflow.com
Curtin University (Australia) programs in occupational safety and health
http://www.curtin.edu.au/curtin/dept/health/ohs/
Curtin University Health and Safety links
http://www.curtin.edu.au/curtin/dept/health/ohs/forum/links/htm
Dave Butler Assoc. http://www.dbainc.com/dba2/library/index.html
Disaster recovery http://www.sgii.com/iw2
Download.com (http://www.download.com)

ESI (project management) http://www.esi-intl.com
Factory Mutual http://www.factorymutual.com
Findlaw http://www.findlaw.com
Global Emergency Management System
 http://www.fema.gov/cgishl/dbml.exe?action=query&template=/gems/g_index.dbm
Happy Puppy http://www.happypuppy.com
Health and Safety UK http://www.healthandsafety.co.uk
Health and Safety New Zealand http://osh.dol.govt.nz/index.html
Hobart Institute of Welding http://www.welding.org
Industry Net Benchmarking Exchange http://www.industry.net/c/services/testmsr
Industry Net (PLCs Online) http://www.industry.net/plcs
ISO 14000 http://www.iso14000.com/
Infraspection Institute http://www.infraspection.com/
Interchange Inc. http://www.interchangeinc.com/
Internet Disaster Information Center (IDIN) http://www.disaster.net
John W's. safety site http://magicnet.net/~johnnw/
LogisticsWeb National Materials Handling Center http://www.logisticsweb.co.uk
Mapquest http://www.mapquest.com
Meynard http://hbmaynard.com
Planning and scheduling benchmarks http://www.neosoft.com/~benchmrx/
Planning and scheduling SIG (UK) http://www.salford.ac.uk/planning/
Polibrid (corrosion and coatings).http://www.corrosion.com
Project Management Forum http://www.pmforum.org
Price watch (computer prices)http://www.pricewatch.com/
RetrieverSoft http://www.retrieversoft.com
Rowan University Management Institute http:plhnet.comPLHCo
Shareware.com http://www.shareware.com
Simul8 http://www.VisualT.com
Truck Net http://www.truck.net/
University of Alabama Continuing Education
  http://bama.ua.edu/~cstudies/
University of Toronto (Business Process Reengineering)
  http://www.ie.utoronto.ca/EIL/tool/BPR.html
University of Wisconsin- Madison School of Business
  http://www.wisc.edu/bschool/erdman
University of Phoenix on-line http://www.uophx.edu/online/
Virtual Reality Institute http://www_ivri.me.uic.edu/
Weather forecast http://cirrus.sprl.umich.edu/wxnet/
Williams Learning Network http://www.willearn.com
Windows95.com http://www.windows95.com
ZDNet Software Library http://www.hotfiles.com

PUBLISHERS AND RETAILERS FOCUSED ON MAINTENANCE ISSUES

Amazon.com http://www.amazon.com
Industrial Press http://www.industrialpress.com
Macmillan Press http://www.mcp.com/
Maintenance Management book list in the UK:
  http://www.conference-communication.co.uk/books/Default.htm

McGraw Hill http://www.mcgraw-hill.com/books.html
Prentice Hall http://www.prenhall.com/
Springfield Resources http://www.maintrainer.com
TWI Press http://www.twi-press.com

RESEARCH SITES
American Society of Mechanical Engineers (ASME)
        http://www-jmd.engr.ucdavis.edu/jmd/
Architecture, Engineering, and Construction (Canada)
        http://ctca.unb.ca/CTCA/sources/
ASME http://mecheng.asme.org
AutomationNET http://www.automationnet.com
Cybertown http://www.cybertown.com/campeng.html
Design Info http://www.designinfo.com
Industrial trade journals (site by AT&T) http://www.ichange.com
Mechanical Engineering Magazine http://www.memagazine.org/index.html
Multi-Media Handbook for Engineering Design http://www.dig.bris.ac.uk/hbook/
NASA abstract and technical report services
        http://techreports.larc.nasa.gov/cgi-bin/ntrs
National Institute of Standards and Technology http://www.nist.gov/welcome.html
National Institute of Standards and Technology Standard Reference Database
        http://www.nist.gov/srd/srd.htm
PartNET company http://part.net
Power Transmission Handbook (excerpts) http://www.hoppenstedt.de/componet
Reliability Analysis Center http://rome.iitri.com/rac/
Rose-Hulman Institute of Technology
        http://www.civeng.carleton.ca/Other_Info/Other_Information.html
Shock and Vibration Analysis Center http://saviac.usae.bah.com
Stanford University Mechanical Engineering Library
        http://cdr.stanford.edu/html/WWW-ME/home.html
Tribology http://www.shef.ac.uk/~mpe/mattrib/tribology
University California Davis Advanced Highway Maintenance
        http://www-ahmct.engr.ucdavis.edu/ahmct/)
University of Maryland National Information Center for Reliability Engineering
        www.glue.umd.edu/enre/reinfo.htm
        Tools page: (http://www.enre.umd.edu/rmp.htm)
University of Massachusetts library
        http://www.ecs.umass.edu/mie/labs/mda/dlib/dlib.html
University of Tennessee Maintenance Technology Laboratory
        http://www.engr.utk.edu/mrc/
Web based information and research resource http://www.eiq.com
        WWW virtual library (Engineering section)
http://www.w3.org/hypertext/DataSources/bySubject/Overview.html
        WWW virtual library on energy
        http://solstice.crest.org/online/virtual-library/Vlib-energy.html

SEARCH SITES
Alta Vista http://www.Altavista.digital.com
DejaNews http://www.dejanews.com/

Dogpile www.dogpile.com
Excite http://www.Excite.com
Galaxy Search http://galaxy.tradewave.com/search.html
Hotbot http://www.Hotbot.com
Infoseek http://www.Infoseek.com
Lycos http://www.Lycos.com
Mamma-Mother of all search engines www.mamma.com
Multi search engine search http://m5.interence.com/ifind/
Open Text Index http://index.opentext.net/
SavvySearch http://www.cs.colostate.edu/~dreiling/smartform.html
Web Crawler http://www.Webcrawler.com
Yahoo http://www.Yahoo.com

SUPER SITES

Frank Measier http://www.efn.org/~franka/MMeasier/TIPS.html
Industrial Link http://www.industrylink.com/
Industry Net http://www.industrynet.com
Industrial Sourcebook http://www.industrialcourcebook.com
Industrial View http://www.industrialview.com
Iron and Steel Institute http://www.steel.org/
Manufacturing and the Internet http://mtiac.hq.iitri.com/mtiac/mfg.htm
Manufacturing Online http://www.chesapk.com/~chesapeake/
Manufacturing Information Net http://mfginfo.com/home.htm
Manufacturing Information Solutions Web site http://conduit2.sils.umich.edu/
Metal Machining and Fabrication International Internet Directory
        http://www.mmf.com/metal/
Mining Internet service http://www.miningusa.com
Plant Maintenance Resource Center
        http://www.iinet.net.au/~sdunn/maintenance/CMMS_vendors.html
Reed Exposition http://www.techexpo.com
Springfield resources http://www.maintrainer.com
Sunbelt Engineering http://www.sunbeltengineering.com/mis_Ink.htm
Sweets http://www.sweets.com/
Thomas Register http://www.thomasregister.com.

Screen Credits

The following Internet Images were used with permission from:

Chapter 2
Page 9   reproduced with permission from AT&T WORLDNET Services

Chapter 3
Page 33   reproduced with permission from Bretech Engineering Ltd.
Page 34   reproduced with permission from Stanford Center for Design Research
Page 34, 69   reproduced with permission from Plant Services/MRO
Page 37   reproduced with permission from Omega
Page 40   reproduced with permission from Yahoo! Inc.
Page 38, 39   reproduced with permission from General Pump
Page 39   reproduced with permission from W.W. Grainger Inc.

Chapter 4
Page 45   reproduced with permission from Yahoo!
Page 60   reproduced with permission from Bretech Engineering Ltd.
Page 61   reproduced with permission from Thomas Register of American
                        Manufacturers
Page 62   reproduced with permission from Industry.net Inc.
Page 63   reproduced with permission from PEM Industrial Sourcebook

Chapter 5
Page 68   reproduced with permission from Plant Engineering
Page 69   reproduced with permission from Plant Services/MRO
Page 72   reproduced with permission from Industrial Press,Inc.
Page 73   reproduced with permission from TWI Press, Inc.
Page 74   reproduced with permission from Association for Facilities Engineering

Chapter 6
Page 77   reproduced with permission from AB SKF, Group Communication
Page 79, 83   reproduced with permission from Asea Brown Boveri Inc.
Page 81   reproduced with permission from American Olean
Page 87   reproduced with permission from Trader Publishing Company
Page 88   reproduced with permission from W.W. Grainger, Inc.

Chapter 7
Page 109   reproduced with permission from Inframetrics Inc.

Chapter 11
Page 154   reproduced with permission from Apollo Group Inc.